D0460944

UNIVERSE

Yolo County Library
226 Buckeye Street
Woodland, Ca 95695

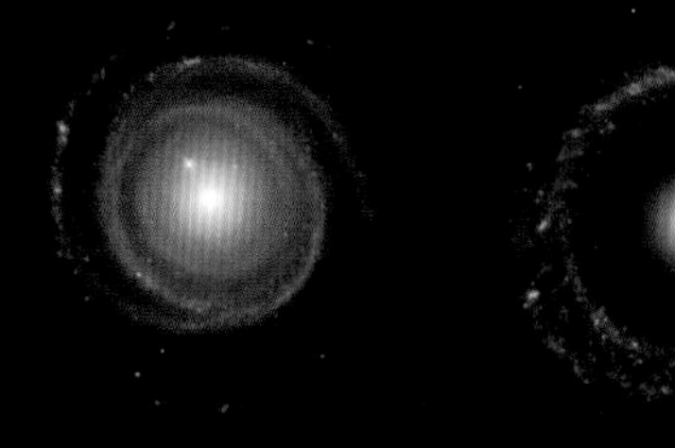

UNIVERSE

A Journey from Earth to the Edge of the Cosmos

Nicolas Cheetham

Smith
Davies

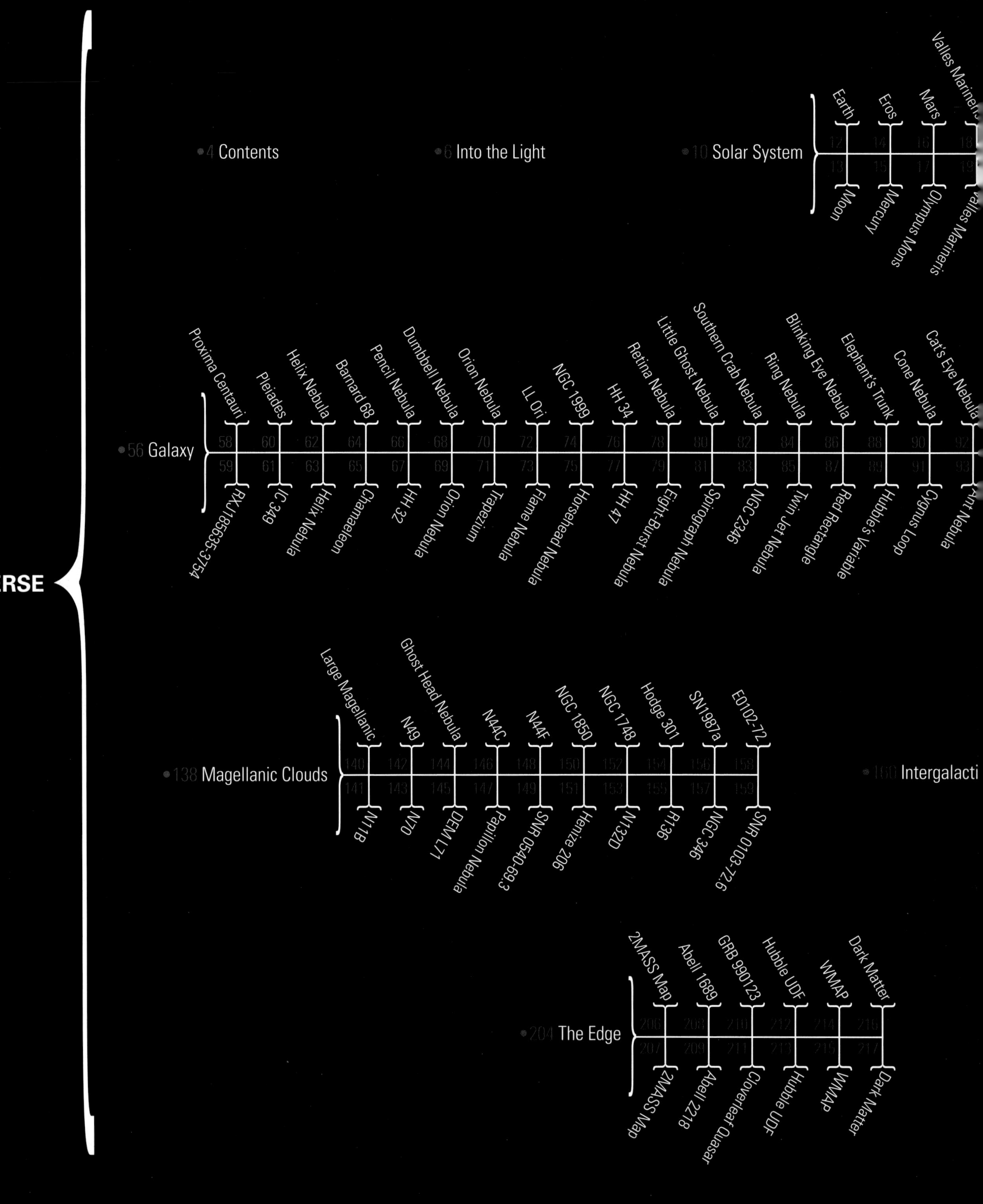

UNIVERSE
4 Contents
6 Into the Light
10 Solar System
Earth
Moon
Eros
Mercury
Mars
Olympus Mons
56 Galaxy
Proxima Centauri
RXJ185635-3754
Pleiades
IC 349
Helix Nebula
Helix Nebula
Barnard 68
Chamaeleon
Pencil Nebula
HH 32
Dumbbell Nebula
Orion Nebula
Orion Nebula
Trapezium
LL Ori
Flame Nebula
NGC 1999
Horsehead Nebula
HH 34
HH 47
Retina Nebula
Eight-Burst Nebula
Little Ghost Nebula
Spirograph Nebula
Southern Crab Nebula
NGC 2346
Ring Nebula
Twin Jet Nebula
Blinking Eye Nebula
Red Rectangle
Elephant's Trunk
Hubble's Variable
Cone Nebula
Cygnus Loop
Cat's Eye Nebula
Ant Nebula
138 Magellanic Clouds
Large Magellanic
N11B
N49
N70
Ghost Head Nebula
DEM L71
N44C
Papillon Nebula
N44F
SNR 0540-69.3
NGC 1850
Henize 206
NGC 1748
N132D
Hodge 301
R136
SN1987a
NGC 346
E0102-72
SNR 0103-72.6
160 Intergalacti
204 The Edge
2MASS Map
2MASS Map
Abell 1689
Abell 2218
GRB 990123
Cloverleaf Quasar
Hubble UDF
Hubble UDF
WMAP
WMAP
Dark Matter
Dark Matter

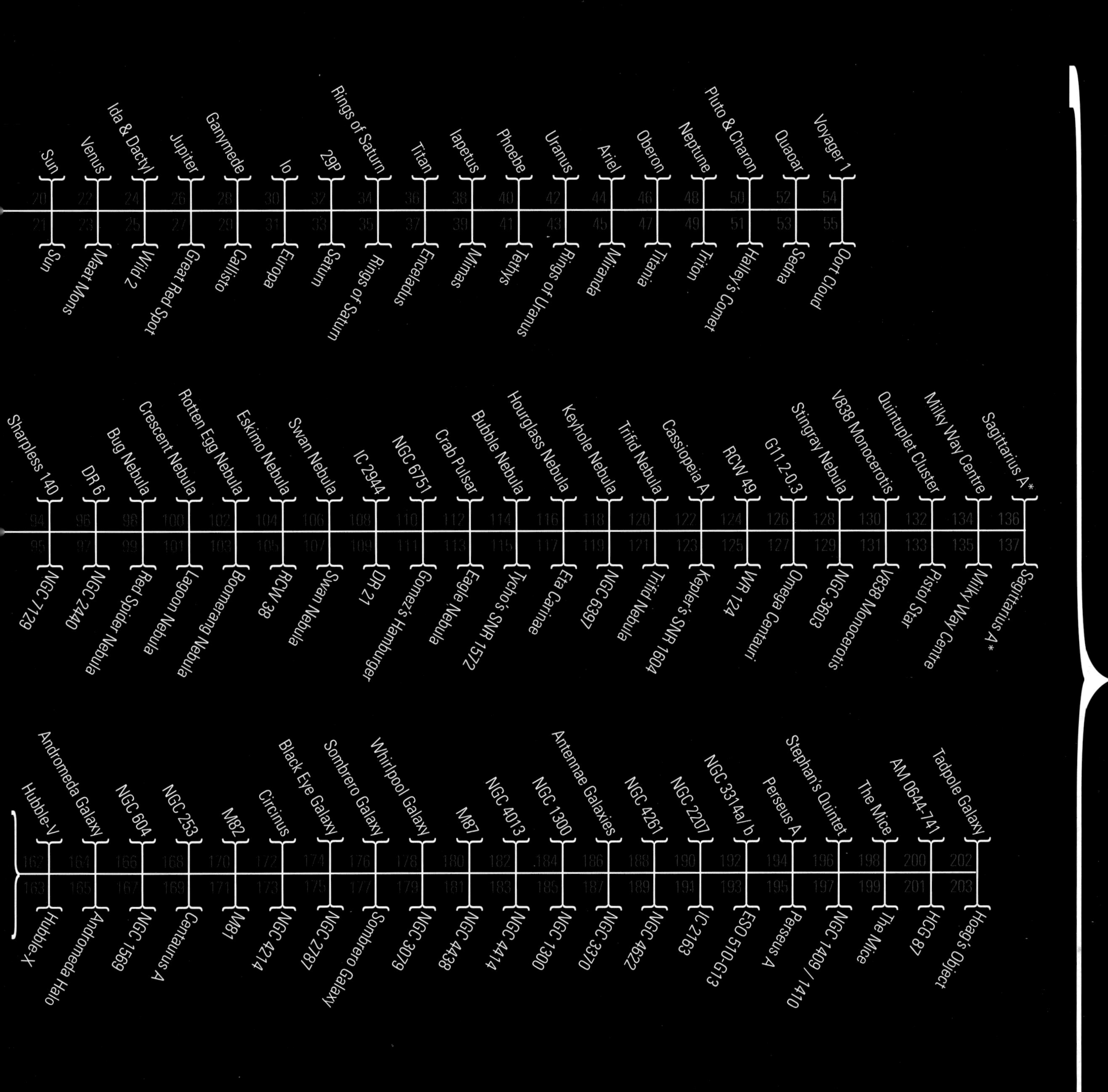

Into the Light

Tonight the light of an ancient supernova will finally reach Earth.

Untold millennia ago, in a distant galactic halo, a giant star spent the last week of its life forging an iron heart. For eleven million years alchemy had fuelled the star's defiance of gravity, its thermonuclear crucibles transmuting first hydrogen into helium, then helium into carbon, carbon into oxygen, oxygen into silicon and finally, silicon into iron. But for a star a ferrous heart is a death sentence – for not even a stellar alchemist can meld iron nuclei together. Metallic heart completed, there was nothing left to burn. Fusion ceased and the supergiant abandoned itself to gravity.

Given free rein, gravity had the might to accomplish what the star could not: the transmutation of iron. Within a tenth of a second the star's core, a sphere the size of our Sun, collapsed into the exotic stuff of neutron stars – a knot of neutronium mere tens of kilometres across. Such an instantaneous metamorphosis unleashed a 21 million kilometer per hour (13 million mile per hour) shockwave and a blaze of multibillion-degree radiation that within a matter of hours had torn open the remains of the star and flooded the heavens with the light of 10 billion Suns.

For millions of years, this luminous conflagration has surged across the universe, carried on a wave of photons travelling at 300,000 kilometres per second (186,000 miles per second). Century by century, decade by decade, year by year, it has relentlessly ploughed towards us. Twelve months ago it breached the Oort Cloud, that cemetery of icy bodies that marks the outer boundary of our solar system, in a matter of hours it will sweep past Voyager 1, mankind's most distant sentinel, and before tomorrow dawns it will finally illuminate Earth's night skies.

But distance will have taken its toll. The supernova will announce itself not as a torrent of light bright enough to read by (like the detonation recorded by Chinese astrologers in 1054 AD) but as a stealthy trickle of radiation, invisible to the naked eye.

This particular stellar catastrophe is, of course, hypothetical, but we can be absolutely certain that photons hailing from a remote supernova are tearing towards us right now and even that they will strike Earth tonight, for somewhere in the observable universe a massive star explodes every few seconds.

Of course, not all photons arriving at Earth have such dramatic origins, and not all traverse millions of light years to reach us, but all photons act as messengers carrying crucial dispatches on the composition of the universe and its energy states – we are bathed in a stream of light carrying both local bulletins and faint

Top left: 28,000 light years away, M80 is one of the densest of the Milky Way's 151 globular clusters. Over 100,000 stars swarm in a volume no more than 100 light years across. In such tightly packed conditions, stellar collisions are not uncommon.

reports from distant shores. Light is the medium through which we experience our cosmic habitat; without it the universe would be a dark and secret place. And when we see, we can understand: the processes that destroyed our imaginary star were divined from the assembled evidence of hundreds of supernovae observations.

Travelling 9.5 trillion kilometres (5.9 trillion miles) a year, there is no swifter messenger for such vital information, but the immense scale of the universe dwarfs even light's velocity, imposing a communication lag between all cosmic events and ourselves. By the time the light of an extragalactic supernova reaches us, entire stellar generations will have passed in its home galaxy and the debris will have been reprocessed inside another star – the original conflagration won't even be a memory. Light's finite velocity means that all roads leading into space also lead into the past. At the expense of never experiencing the entire universe as it exists at this instant, we can directly explore its deep past. Simply by looking deep into the universe we become eyewitnesses to its evolution, able to reconstruct its history. Ultimately, such a view of the cosmos is infinitely more valuable than a complete view of its current state.

Despite being suffused with light, surprisingly little of the universe is visible to the naked eye. Unaided, we can detect only 6,000 of the universe's estimated 70 billion trillion stars from Earth – and no more than a few thousand from any one location. Dust, distance, light pollution and a thick atmosphere may serve to obliterate much of this population, but the simple truth is that we are blind to most wavelengths of light and subsequently blind to a large proportion of the celestial sphere's denizens.

Beyond the familiar rainbow of visible light there lies a much more vibrant world, for the heavens blaze across a spectrum of energies far broader than the eye can behold: black holes are uncloaked by bright x-ray bursts as matter spirals into them; treasure chests of embryonic stars smoulder in the infrared; nebulae bask in the ultraviolet light of hypergiant stars; and microwave echoes chronicle the afterglow of the Big Bang.

Only in the last century have we become aware of light's many guises and been able to step into this unseen realm – it is no coincidence that our exploration of space has advanced hand in hand with our exploration of the electromagnetic spectrum. Without the technological supersenses to capture the invisible – and amplify the visible – our view of the universe would be but a pale reflection of its true nature and our understanding of it similarly constrained.

Our celestial models have always been constrained by the limits of observation: as we step from solar system to Milky Way, and from Milky Way to intergalactic space, we leap the bounds of a number of historical universes. For the astronomers of the ancient world the solar system was the universe. Their cosmos was centred on the Earth and enclosed within a crystal sphere speckled with fixed stars. By 1543, Nicolas Copernicus had reorganized and enlarged this miniature universe, but he failed to demolish its crystal walls. In 1671 Giovanni Domenico Cassini established the distance between the Earth and the Sun and the true scale of the solar system began to emerge. By now astronomers suspected the universe extended far beyond the bounds of the solar system, but they needed to prove it. The transits of Venus in 1761 and 1769 enabled the size of the Earth's orbit to be precisely mapped, which provided the baseline needed to triangulate the distance to the nearest stars. The first accurate stellar parallax (ten light years to 61 Cygni) was fixed in 1838 and our universe took on galactic dimensions.

By the 1920s observation had pushed the bounds of our universe out to some 300,000 light years, enveloping the Milky Way and the Magellanic Clouds but extending no further and containing no other galaxies. The intergalactic breakthrough came in 1924 when Edwin Hubble managed to resolve individual stars in M31, the Andromeda 'nebula', using the cutting-edge 100-inch Hooker Telescope at the Mount Wilson Observatory. From his observations he was able to determine that these stars lay far beyond the most distant fringes of our galaxy. It was clear that the Andromeda 'nebula' was a galaxy in its own right, and if it was, then it stood to reason that other spiral nebulae were too.

By banishing such nebulae from the Milky Way, Hubble had unlocked a night sky that could boast billion light year vistas, but his most important discovery was still to come. By 1929, analysis of these 'new' galaxies had revealed that almost all of them were receding at rates directly proportional to their distance. This relationship implied the universe was expanding and had a definite origin at a precise moment in time – observations that matched the predictions of Einstein's theory of general relativity. It is on this foundation stone that the current Big Bang model of the universe is laid.

By the mid-twentieth century the planets and satellites of the solar system had been comprehensively freed of their crystal shackles and their place in the grander cosmic scheme accurately determined, but observation from Earth had been able to supply little more detail on our close companions. Any atlas of the solar system left vast territories unmapped, an empty canvas where an inscription reading 'here be dragons' would not have been out of place. In the last four decades our robotic emissaries have largely banished these dragons (unless they be subterranean dragons) and replaced them with a rich harvest of imagery and information that cries instead 'here be water' or 'here be methane'. Every planet bar Pluto has been visited, if only in passing.

So, at the dawn of the twenty-first century, having ascended a mountain built from observation, theory and technical expertise, we find ourselves bathed in the light of a celestial panorama that extends for 130 billion trillion kilometres (80 billion trillion miles) in every direction. Only recently unveiled, the scene before us still begs innumerable questions, but right now, it is enough to set such concerns aside and simply enjoy the view: our journey into the universe starts here.

All journeys begin with a single step, and ours is no exception. We cross the 385,000 kilometres (240,000 miles) between Earth and the Moon in a stride. It took the Apollo astronauts four days to make the same voyage, but our journey is at the speed of light and takes only 1.3 seconds. At this velocity Mercury, Venus, Mars and the Sun are minutes distant and we stand only hours from the swarm of icy planetoids that cluster at the edge of the solar system.

To bridge the four years of travel that separate us from our next landmark, Proxima Centauri, we must lengthen our stride, and then lengthen it again as we press on into the Milky Way. Distances are now measured in hundreds and even thousands of light years. On this scale, the rhythms of stellar life unfold before our eyes: we pass through dark nebulae afire with newly smelted stars, watch dying stars bloom and fade and skirt the debris clouds of supernovae. Navigating through thick swarms of stars orbiting mere light weeks from each other we approach the galactic core, a gravitational court of white dwarfs, neutron stars and hypergiants in the thrall of a three million solar mass supermassive black hole.

To leap beyond our galaxy's bounds, we must once more increase our stride. Hundreds of thousands of light years must be dispatched to reach the Milky Way's companion galaxies, the Large and Small Magellanic Clouds, and millions more devoured before we can confidently cross true intergalactic space. Despite the distances between galaxies, intergalactic space is far from empty; it is, in fact, more densely populated by galaxies than interstellar space is by stars. Out here galaxies are not island universes splendid in their isolation but form interacting, merging and evolving flocks. Deep space resounds to their clash.

We now push towards the hazy limits of our vision, a realm where our perspective encompasses billions of light years. Here we can see the large-scale structure of the universal landscape. Massive conglomerations of galaxies cluster like grains of dust to a celestial veil of cobwebs, warping space with their gravity and projecting kaleidoscopic bubbles of primeval light into our present. With hypersensitive infrared eyes we pierce the once barren 'redshift desert' and discover a Burgess Shale of eccentric galactic infants, their bizarre forms witness to an era where order was just emerging from chaos. Another step, and our journey into the light plunges us into the dark: we have entered the cosmic Dark Ages, a universe without stars, without starlight. Then, at a distance of 13.4 billion light years and a mere 379,000 years from the Big Bang, a sudden lightning flash of radiation etches an afterimage of the Big Bang across the heavens. This is our final horizon – we have reached the far shores of light. Beyond this point a dense fog of superheated plasma impedes light's progress: it can take us no further.

In the dark, some of the questions we refrained from asking while we took the view return to haunt us: despite the luminous opulence of the universe, something is missing. The shape, distribution and movement of galaxies cannot be explained unless they are nearly ten times more massive than they appear – we have explored 13.7 billion light years of space and time and yet have seen no sign of this missing mass. If our calculations are correct, this 'dark' matter makes up 87 percent of the universe and everything – planets, stars and galaxies – we have encountered on our journey is but flotsam on its lightless ocean. Ultimately, dark matter is the real stuff of the universe, binding galaxies together and weaving the web to which they cluster. That we have deduced its existence proves that despite remaining largely invisible and mostly intangible, the universe is by no means unknowable.

Containing nearly 200 colour images, *Universe: A Journey from Earth to the Edge of the Cosmos* reveals near space as seen through the eyes of a fleet of satellites and interplanetary probes: Cassini, Deep Space 1, Galileo, Magellan, Mariner, Mars Global Surveyor, NEAR Shoemaker, Stardust, SOHO, TRACE, Viking, and Voyagers I & II. Longer range observations come courtesy of the Hubble Space Telescope, the Chandra X-ray Observatory and the Spitzer Space Telescope. All these images are, of course, much more than celestial snapshots, each one has deepened our understanding of the universe. They have revealed the aftermath of the Big Bang, and confirmed the existence of black holes. The curvature of spacetime as light is distorted around massive galactic clusters provides a graphic proof of one of relativity's least intuitive tenets. Some of the images, for example the 'pillars of creation' in the Eagle Nebula, have become as iconic of deep space as the blue-green globe is of Earth. There can be no doubt these images have coloured our perception of space – sometimes quite literally.

Viewing from the porthole of a passing spaceship, we might be underwhelmed by the ghostly manifestation of many nebulae compared to the bold images that saturate this book. We now know that visible light is but a brief and sparsely populated archipelago rising from an ocean of hidden wavelengths, but even here our vision is limited. The wash of fluorescing gases that creates a nebular cloudscape is often too faint to be properly appreciated by the human eye. Conversely, if we were to travel to Betelgeuse, the red giant, it would burn white, not red, as our eyes were overwhelmed by its radiance.

If only faintly visible, the celestial landscape, just like the terrestrial, can be enhanced by the use of filters and long exposures but the much larger invisible universe can only be revealed by a process of translation, or more accurately, transposition. When an image falls outside the visible spectrum it has to be mapped to optical wavelengths. Because x-rays, radio waves, the ultraviolet and infrared have no colour, the initial colour choice is arbitrary, but it forms the key by which all subsequent wavelengths are transposed: shorter wavelengths are bluer; longer wavelengths are redder. Sometimes even visible light is remapped to enhance otherwise subdued features. The iconic pillars of the Eagle Nebula soar four light years high against a turquoise hydrogen sky. In actual fact fluorescing hydrogen emits a red light – but so does sulphur. To avoid confusion hydrogen, with a shorter emission wavelength, was transposed a few octaves higher into green.

Some have likened such scientific post-production to the modern counterpart of nineteenth century Romantic landscape painting – a careful fusion of art and reality; others simply label it misleading. But just how can one judge the visual accuracy of the invisible?

We need to remember that seeing takes place more in the brain that in the eyeball. Electromagnetic wavelengths are absolute and can be measured precisely, but only after they have been filtered through a mesh of chemicals and electrical impulses inside our heads do they conjure colour. As a result no observation can ever be entirely objective and utterly dispassionate.

The sensitivity of our instruments is staggering. In the 1980s Carl Sagan estimated that the total energy received by all the radio telescopes in the world is less than a single snowflake touching the ground. But just as the biological Mark I Eyeball has its limits, so too does our imaging technology. At optical or near optical wavelengths, discoveries can be distant or small but rarely both. For example, the most distant object discovered in the solar system is the planetoid Sedna at over 13 billion kilometres (8 billion miles) away. But at less than 1800 kilometres (1110 miles) across, Sedna is too small for even the Hubble Space Telescope to resolve clearly and it appears as a blocky clump of pixels. The crude image of Sedna compares unfavourably with the Hubble Space Telescope's crisp image of The Mice, two colliding galaxies nearly three billion trillion kilometres (two billion trillion miles) away. However, The Mice span a distance measured in hundreds of thousands of light years compared with Sedna's mere kilometres, making the Sedna image the more noteworthy optical feat. Resolving Sedna has been likened to trying to spot a soccer ball 1500 kilometres (900 miles) away and represents the current limit of our technical acumen.

Another artefact you will encounter is a distinctive jagged edge that staircases its way across some images. This is a hallmark of the Hubble Space Telescope's Wide Field and Planetary Camera 2 (WFPC2) whose four digital sensors are arranged in a mosaic. As the camera's name suggests, the sensors divide into three wide-field chips and one higher-resolution central 'planetary' chip.

Accompanying every image are three boxes. The first gives the distance, in light seconds, minutes or years to the object from the chair you are sitting on. Of course we live in a dynamic universe and everything is moving, so for the sake of convenience the distance to all solar system objects is fixed to the positions they occupied on 1 July 2005. On the galactic scale the whole solar system churns around the hub of the Milky Way, which is itself rushing towards Andromeda while just about everything else rushes away from it, but as the timescale of these changes runs into millions of years any distance updates can wait for the second edition of this publication.

The second box contains the name of the object pictured and a brief note identifying what it actually is, be it planet, pulsar or proto-planetary nebular. Not all names trip off the tongue. We have struggled to keep up with the rush of celestial bodies astronomy has brought us. The nine major planets rejoice in the names of Greek and Roman deities. Their satellites form an ensemble cast of titans, muses, faeries and great Shakespearean roles. At the edges of the solar system we encounter Sedna, an Inuit goddess of the underworld, and Quaoar, a Native American creation deity. Beyond this point numbers begin to replace names.

Like entomologists collecting insects, astronomers have amassed stars and galaxies, grouping them according to a multitude of taxonomies – over 2,000 separate cataloguing systems now exist. The first catalogue was compiled by Charles Messier between 1758 and 1781 and contains 105 'nebulous' objects. In deference to Messier's efforts any object that appears in his catalogue is listed here under the designation M followed by its index number. The letters NGC stand at the head of most of the entries in this book, standing for New General Catalogue (NGC). 'New' is a relative term since the catalogue was first published in 1888. It contains 7800 visible non-stellar bodies (clusters, galaxies, nebulae) and forms the 'standard' list of celestial objects. In 1895 and 1908 it was expanded by two Index Catalogue (IC) supplements. More recently discovered items tend to be catalogued by their position in the sky.

The final box contains a brief caption that adds some background information to the image – any unfamiliar terms should be defined in the short glossary (pages 218–220). The captions can be read in any order, but taken sequentially they form a journey from Earth to the very edge of the cosmos. To begin the Grand Tour, just turn the page...

1
the solar system

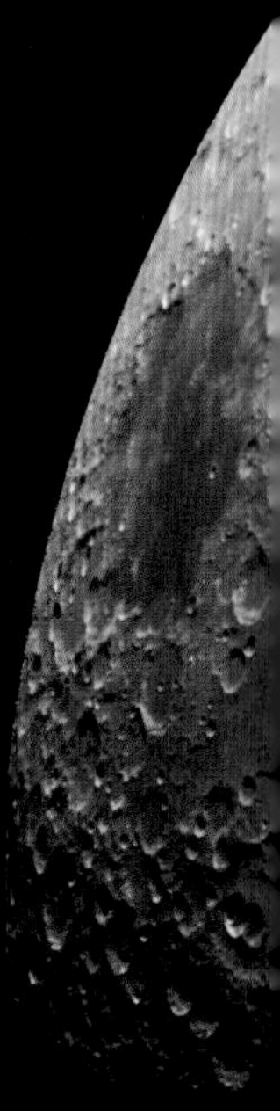

0 light distance	*Earth* Planet	Our tenure on the universe's largest known rocky planet is as brief as our habitat is narrow: excluded from two-thirds of the planet by the deep ocean, the surface of Mars is better mapped; and what we consider truly Earth-like conditions will prevail for less than ten percent of the planet's overall lifespan. With this in mind, there is no reason to delay our departure for the universe.

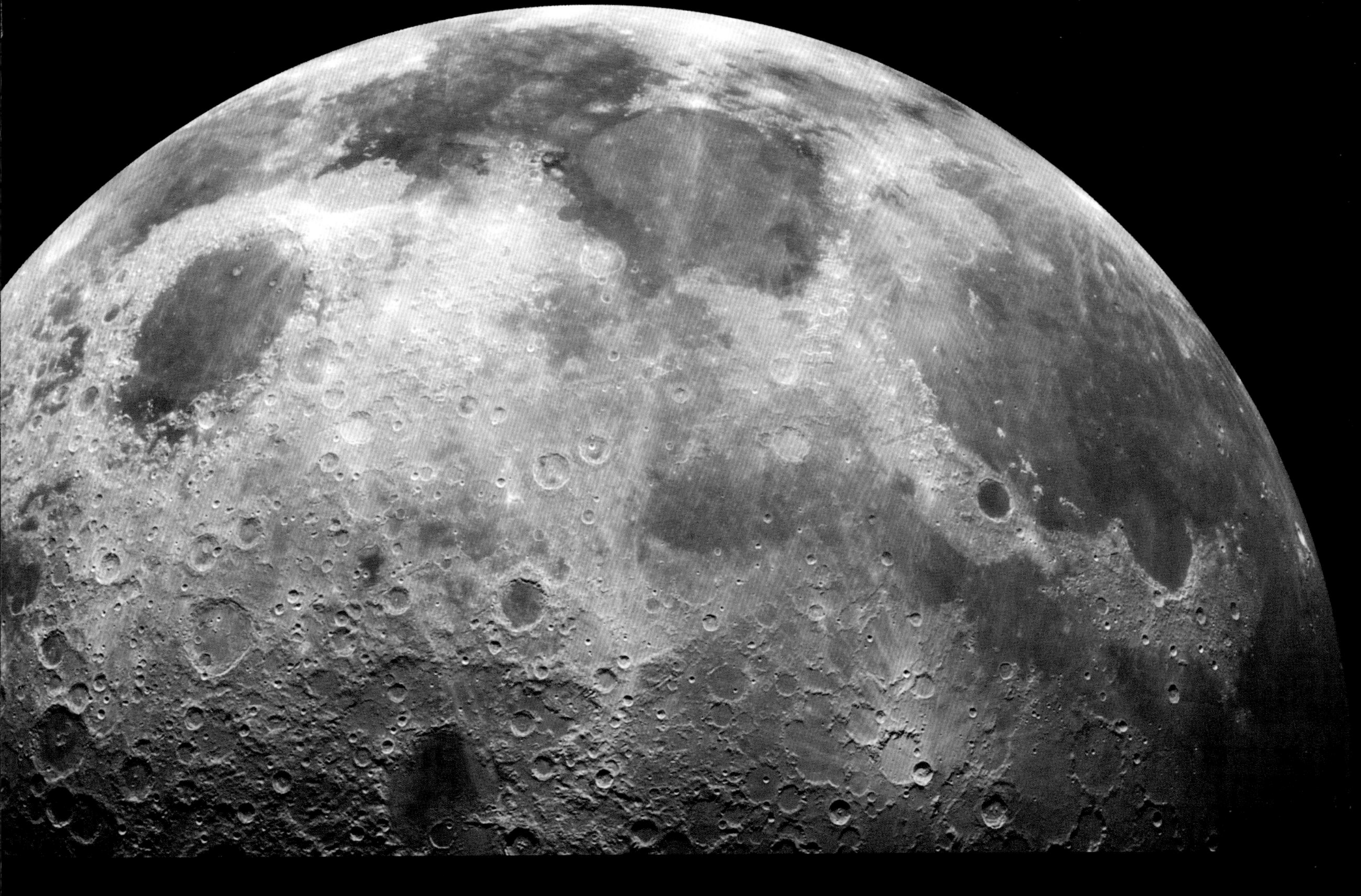

1.3	*Moon*	Only 1.3 seconds into our voyage and we have already reached the limits of human exploration, for the Moon is mankind's only footfall on an alien world. In some respects, the Earth and Moon could be considered a binary planet system: in comparison to its parent, the Moon is by far the largest satellite body in the solar system.
light seconds	Moon	

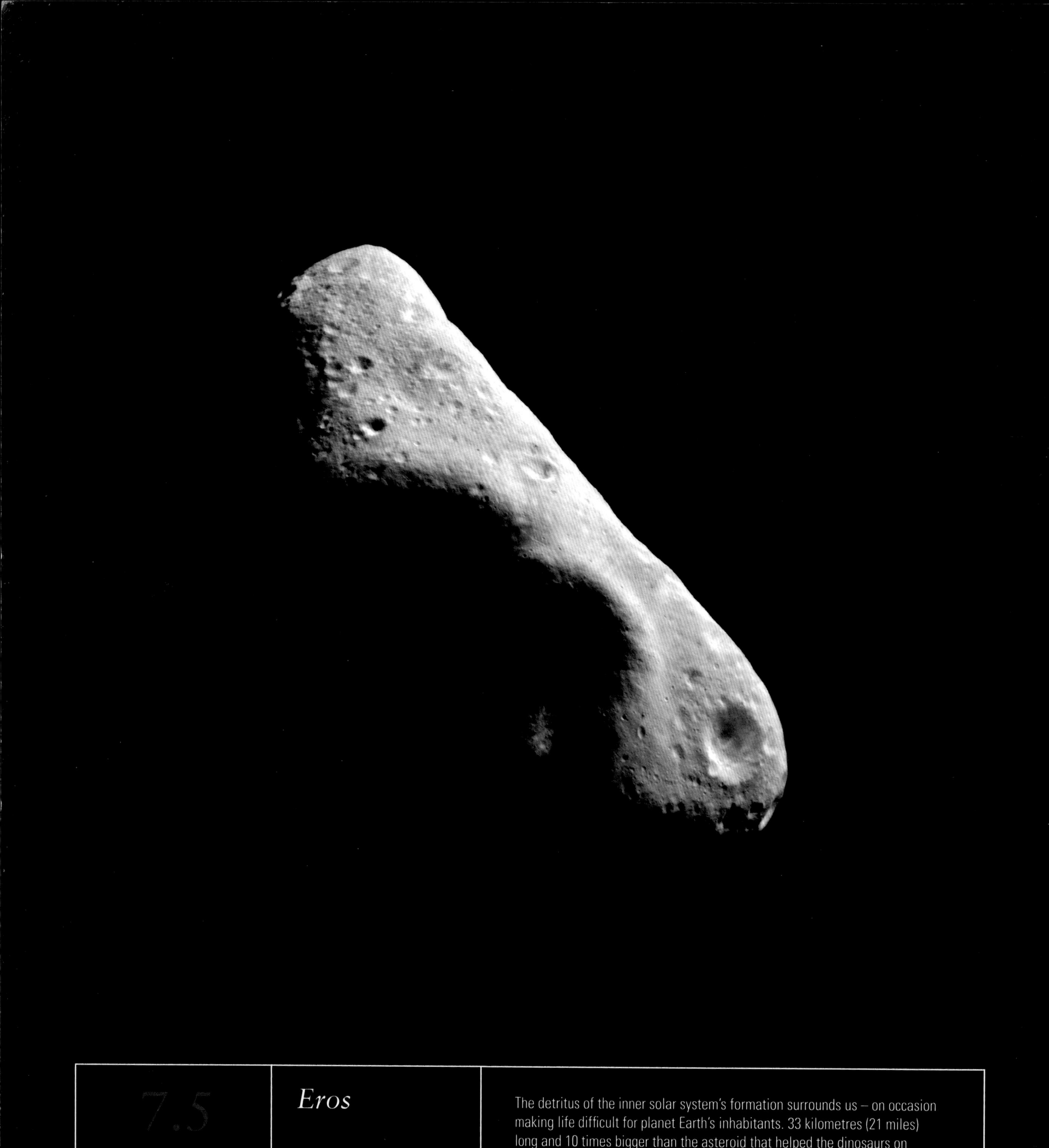

7.5 light minutes	*Eros* Asteroid	The detritus of the inner solar system's formation surrounds us – on occasion making life difficult for planet Earth's inhabitants. 33 kilometres (21 miles) long and 10 times bigger than the asteroid that helped the dinosaurs on their way, Eros is, thankfully, a well-behaved neighbour. Despite at times approaching to within 75 light seconds of Earth, its orbit never crosses ours.

8.1 light minutes	*Mercury* Planet	Mercury has the widest temperature range of any planet in the solar system, from -170 °C (-274 °F) at night to 350 °C (662 °F) during the day. Its landscape is dominated by the Caloris Basin, the 1,300 kilometre- (800 mile-) wide legacy of an impact that remodelled the planet, sculpting a global web of concentric mountain ranges, lava plains, ridges, faults and depressions.

8.3

light minutes

Mars

Planet

With summer temperatures reaching 27 °C (80 °F), Mars has the kindest climate of our planetary neighbours. Without a spacesuit you could survive here for tens of seconds. On Mercury you wouldn't survive a single second and on Venus not even a spacesuit would help you. Despite such clement conditions, the jury is still out over whether Mars has ever harboured life.

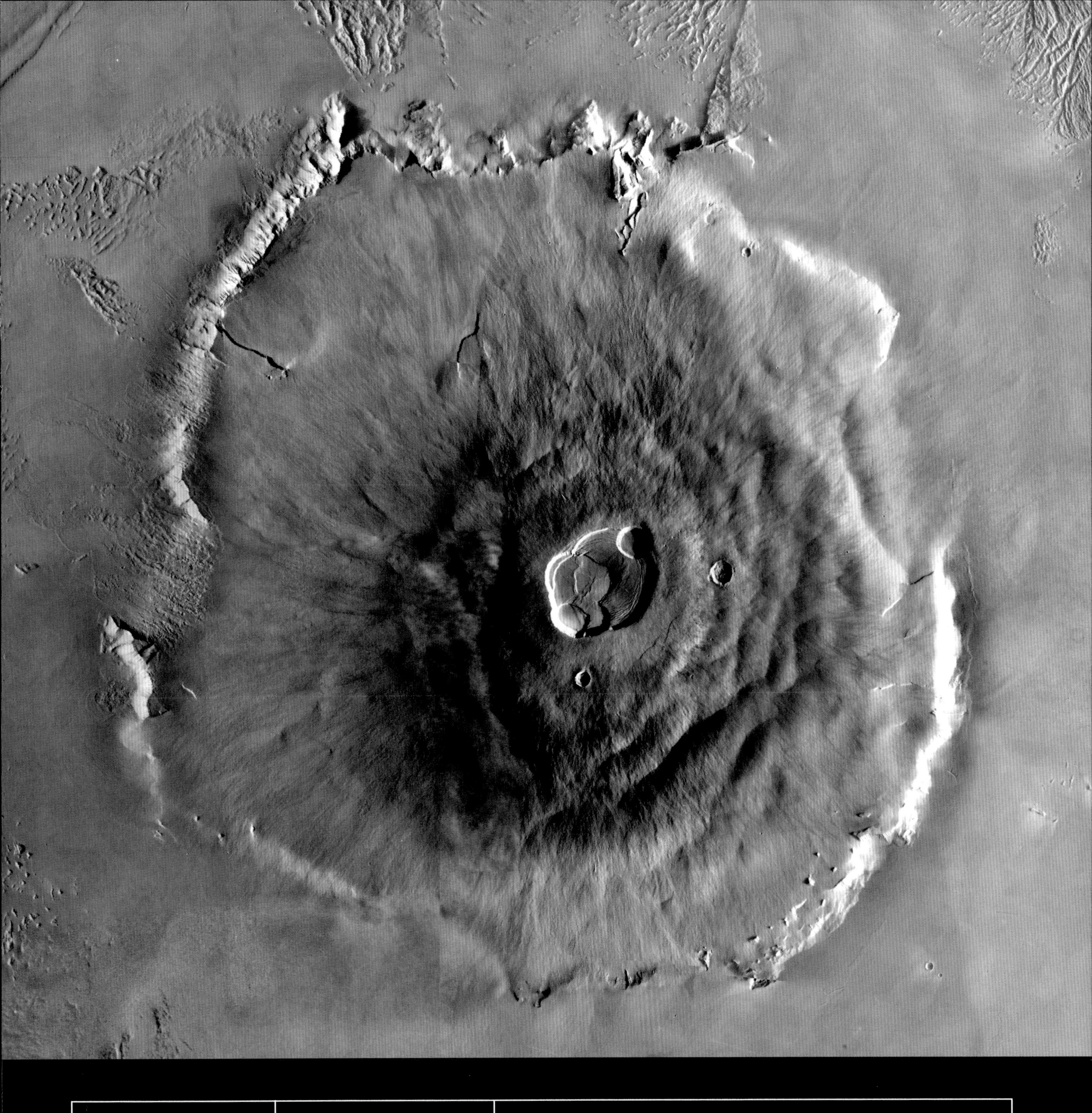

8.3

light minutes

Mars

Olympus Mons

Volcano

Mars' scenery is altogther more monumental than Earth's. Towering 24 kilometres (15 miles) above the surrounding plain, Olympus Mons is the solar system's largest volcano. Its base is over 600 kilometres (375 miles) in diameter and is rimmed by a 6 kilometre (3.75 mile) high escarpment. Such geological exuberance is possible thanks to low gravity and a lack of tectonic motion.

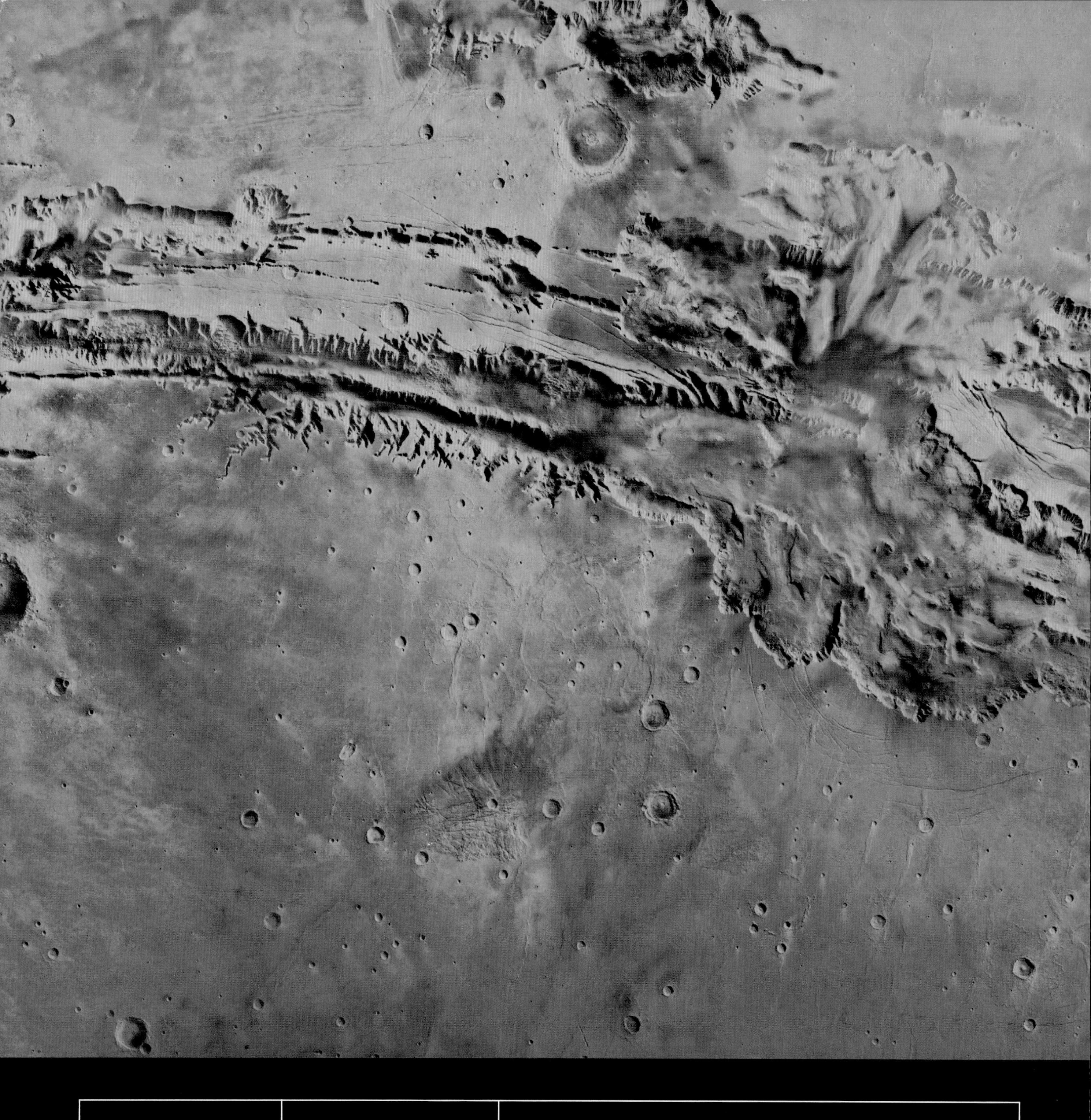

8.3 light minutes	*Mars* Valles Marineris Rift valley	Lying at the base of cliffs that plunge 6 kilometres (3.75 miles) and reaching almost a quarter of the way around the planet, Valles Marineris is the Grand Canyon of Mars. Despite increasing evidence that water may once have flowed on the planet, this fissure was not formed by running water; it is a tectonic stretch mark created by the growth of nearby massive volcanoes.

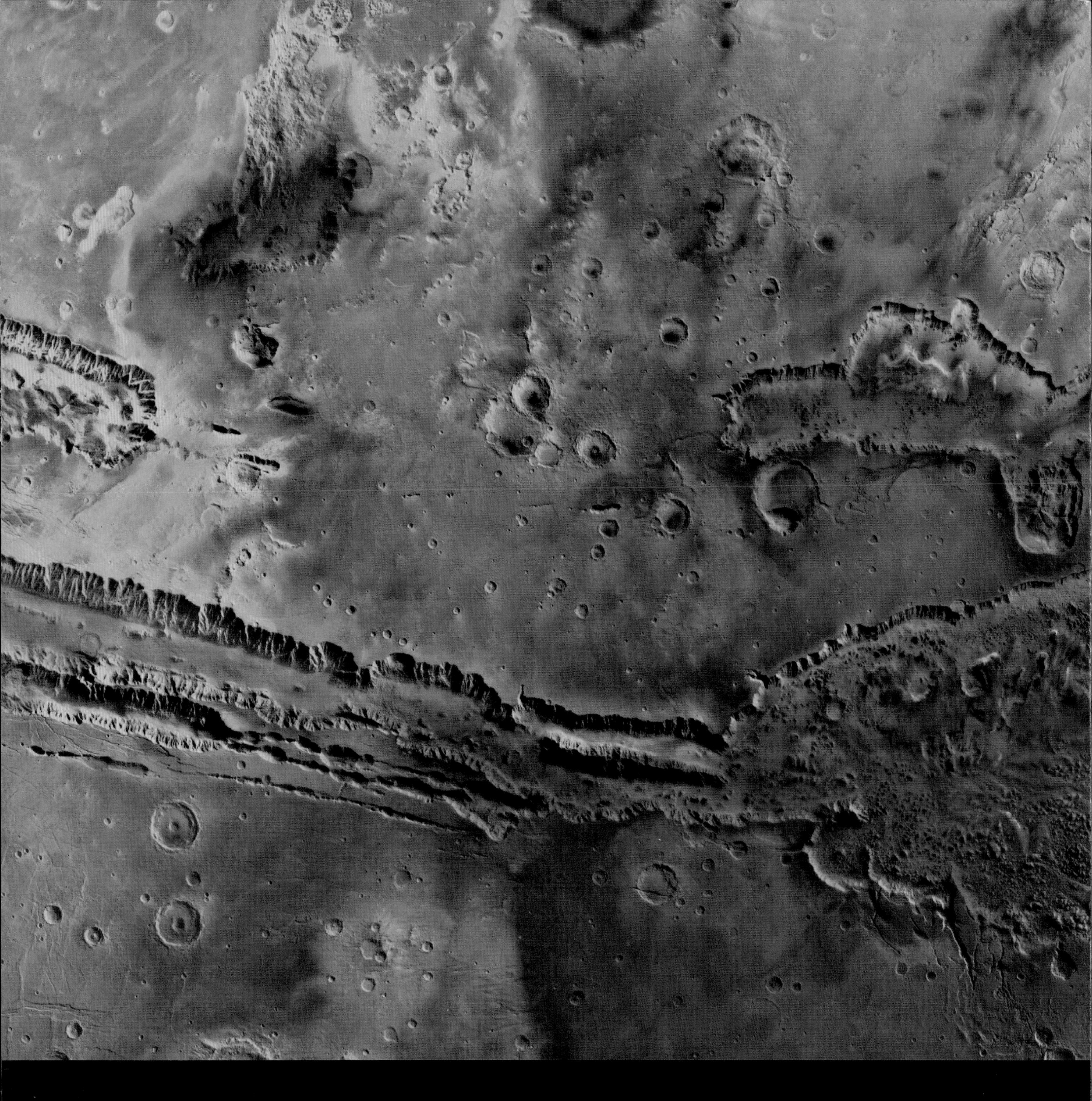

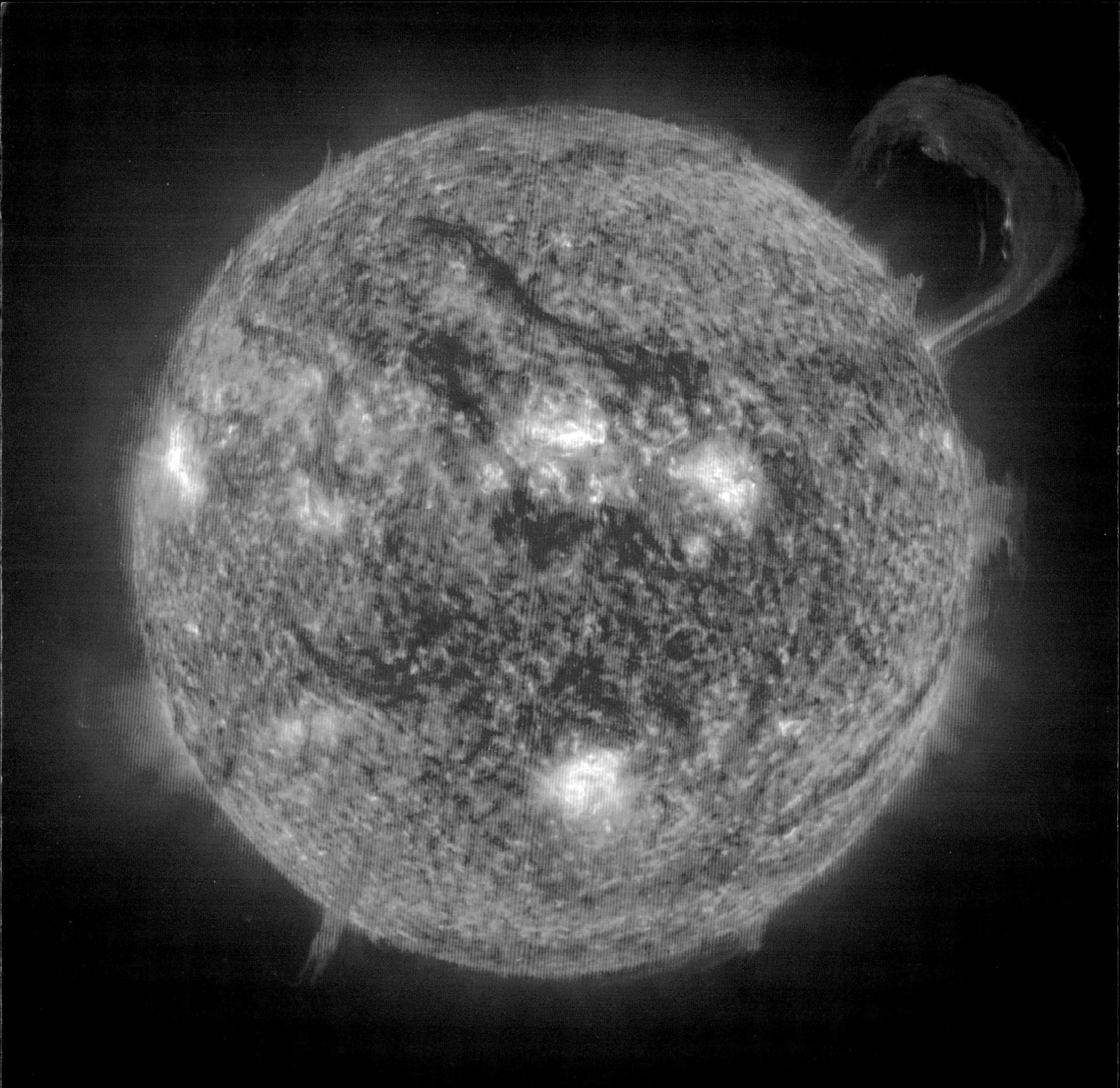

8.5 light minutes	*Sun* Star	One of 100 billion stars in our galaxy, our sun is an unremarkable, middle-aged yellow dwarf orbiting the galactic core once every 225 million years. At 1.4 million kilometres across (864,000 miles) it contains 99.8 percent of our solar system's mass, which it burns at a rate of five million tonnes every second, producing 383 billion billion megawatts of energy.

8.5 light minutes	*Sun* Solar flare	With the energy of up to a billion megatons of TNT, solar flares can hurl clouds of plasma into space at velocities approaching the speed of light and cover a region the size of many Earths. This eruption measured over 100,000 kilometres (62,000 miles) in height. Solar flares are believed to result from the violent release of the magnetic energy from the area surrounding sunspots.

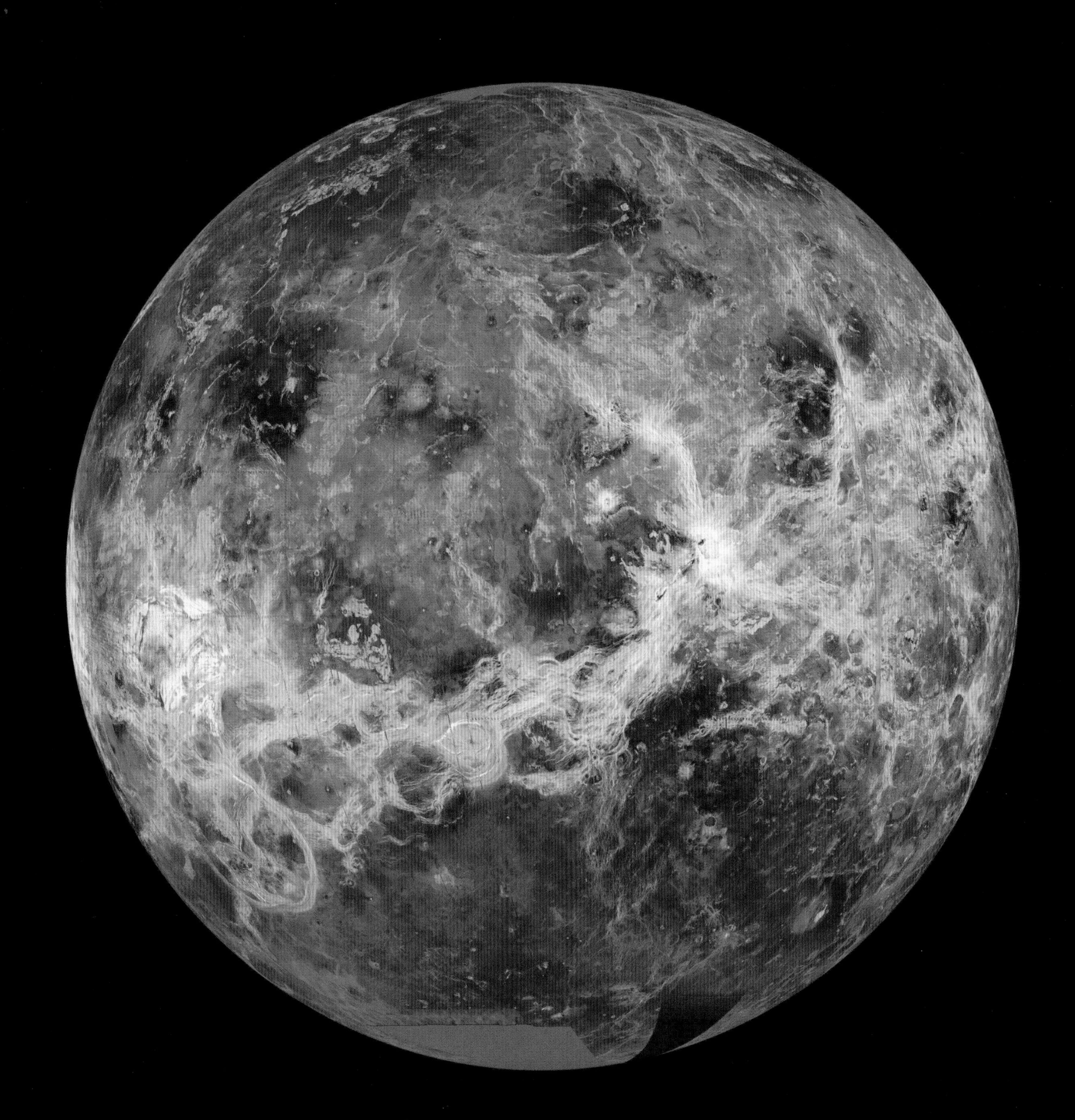

12.6 light minutes	*Venus* Planet	Superficially, Venus resembles Earth with a similar mass, size and density. In reality, it could not be more different. Wrapped in a cloak of carbon dioxide laced with sulphuric acid, its surface bakes at over 400 °C (750 °F) under an atmospheric pressure 90 times greater than that of our planet. Not an ideal location for a day trip – especially as a Venusian day lasts 243 Earth days.

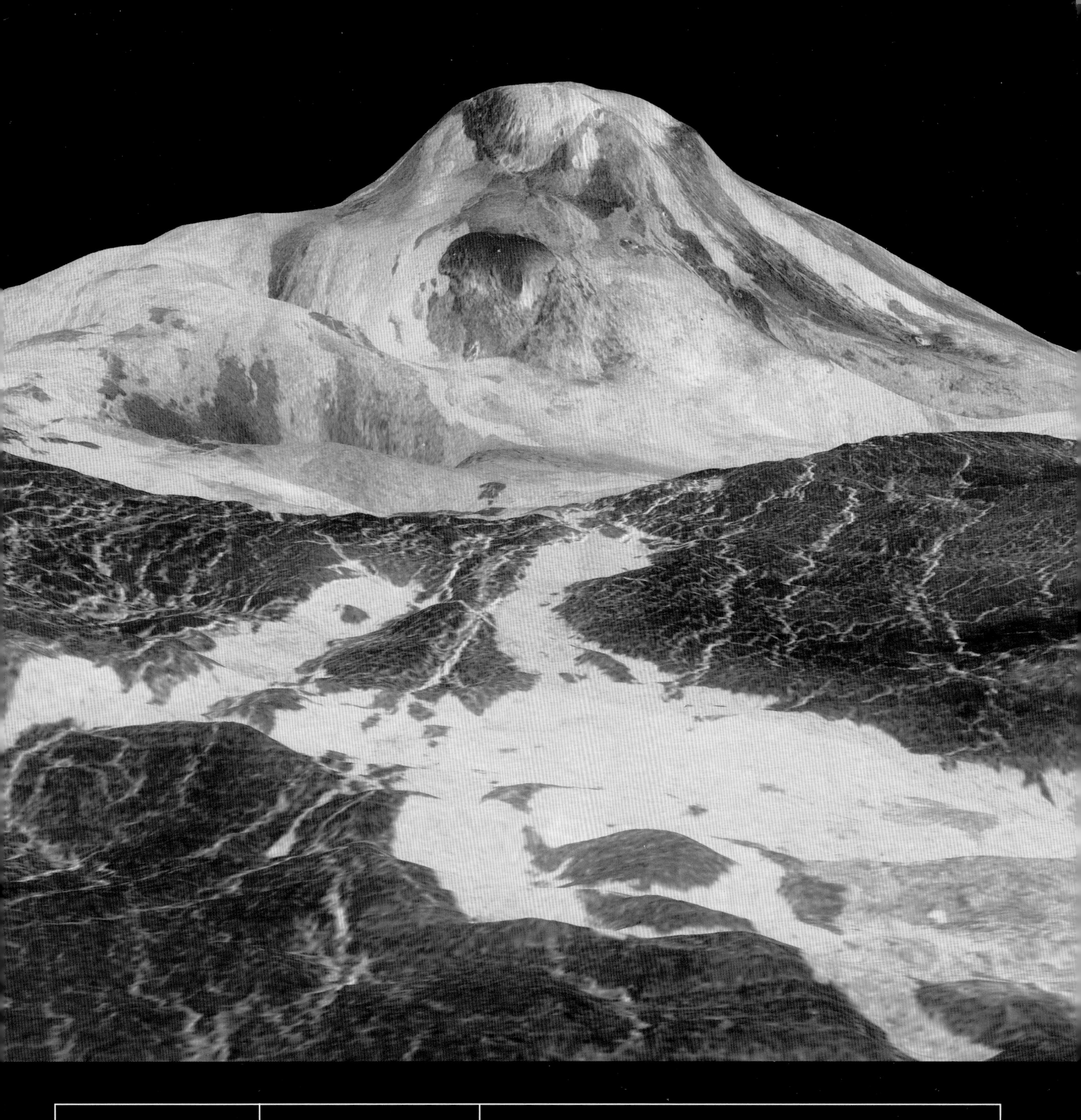

12.6 light minutes	*Venus* Maat Mons Volcano	Under the blistering clouds of noxious gases, Venus is a planet of rolling plains punctuated by volcanic massifs. Maat Mons, rising to 8 kilometres (5 miles), is Venus' second highest peak. It is a young landscape too, apparently completely resurfaced 500 million years ago by a massive episode of global vulcanism. No wonder Venus has been described as a 'planet-wide catastrophe'.

24.2

light minutes

Ida & Dactyl

Asteroids

Orbiting between Mars and Jupiter, the Asteroid Belt marks the divison between the rocky inner and the gaseous and icy outer solar system. Considered to be the fragments of a stillborn planet, asteroids range in size from mere dust particles to 900 kilometre (500 mile) behemoths. This belt resident, Ida, is 52 kilometres (32 miles) long and accompanied by its own satellite, Dactyl.

32.2 light minutes	*Wild 2* Comet	Prior to 1974, Wild 2 was a denizen of the outer solar system, patrolling the gas giants before a particularly close encounter with Jupiter transferred its beat to the asteroid belt. Five kilometres (three miles) across and composed of frozen water, other ices, organic materials and fragments of rock, Wild 2 will unfurl a distinctive cometary tail when it passes within 25 light minutes of the Sun.

44.5 light minutes	*Jupiter* Planet	Jupiter is often described as a failed star – it is made of the right ingredients but would need to be at least 80 times more massive to make the grade. A more positive observation is that our solar system's colossus is about as large in diameter as a gas giant can be. If you fed extra material into the planet it would be compressed by gravity rather than significantly adding to Jupiter's girth.

44.5 light minutes	*Jupiter* Great Red Spot Storm	At 25,000 kilometres (15,500 miles) long and half as wide, Jupiter's most famous feature is large enough to engulf Earth twice over. Having raged for at least three and a half centuries the Great Red Spot could be regarded as the perfect storm. Fuelled by heat from Jupiter's core, it endures because there are no landmasses to dissipate its energy, unlike its hurricane cousins on Earth.

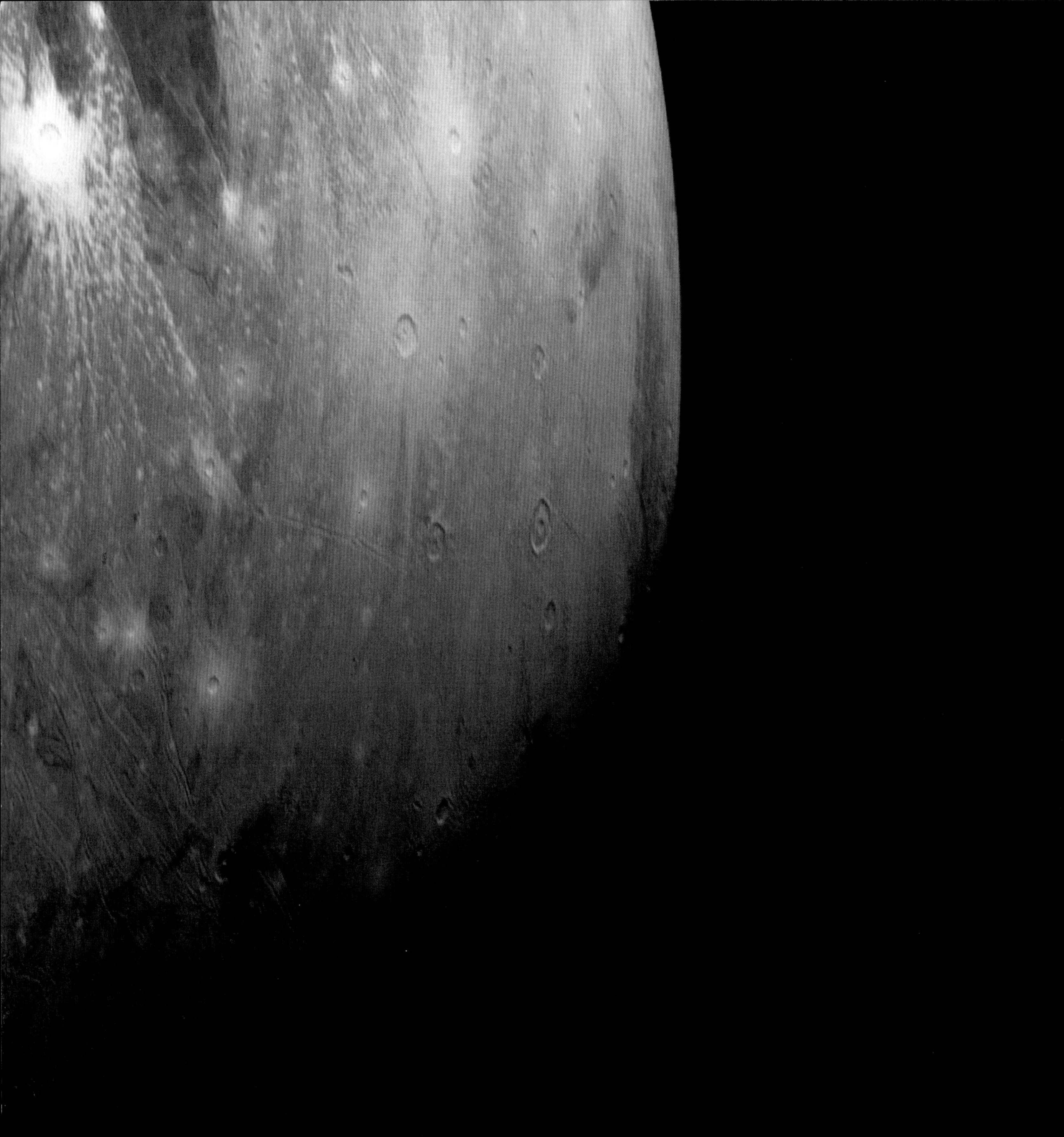

44.5	*Ganymede*	With 63 satellites, Jupiter is surrounded by its own solar system. Apart from the four large Galilean moons, only a handful measure more than a few kilometres across. Ganymede, with a diameter of 5,262 kilometres (3,280 miles), is larger than Mercury (a distinction it shares with Saturn's Titan) and the largest moon in the solar system.
light minutes	Moon	

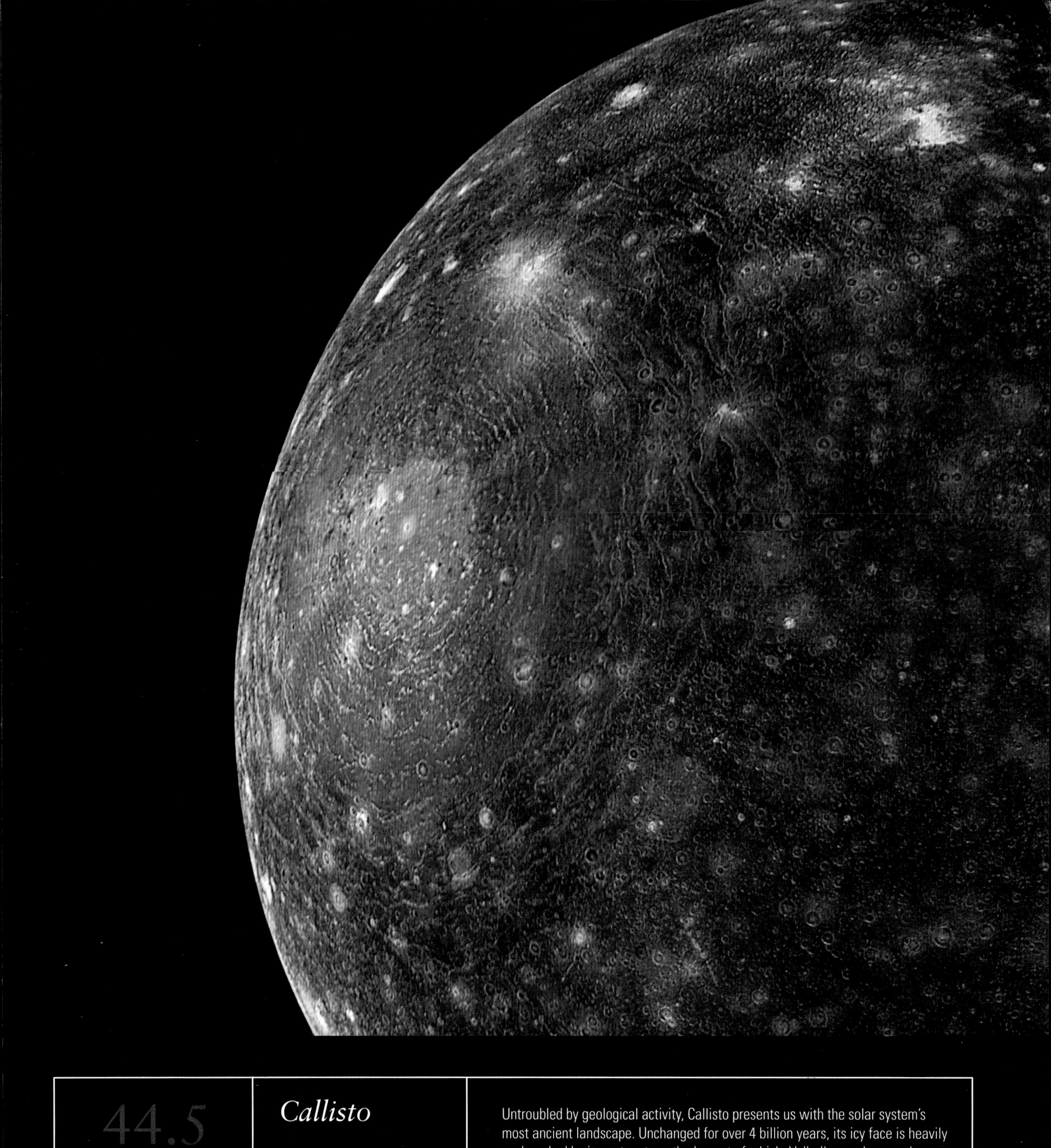

44.5

light minutes

Callisto

Moon

Untroubled by geological activity, Callisto presents us with the solar system's most ancient landscape. Unchanged for over 4 billion years, its icy face is heavily pockmarked by impact craters, the largest of which, Valhalla, can be seen here. The impact basin is 600 kilometres (375 miles) across, and is surrounded by concentric ripples that extend over a third of the way around the moon.

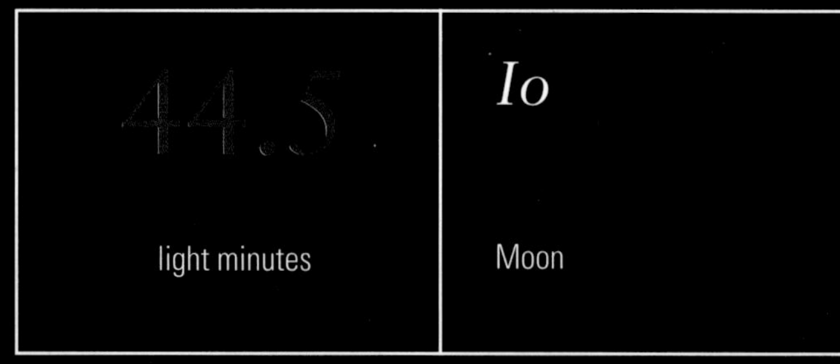

44.5

light minutes

Io

Moon

Tormented by the gravitational tides of its sister moons and its parent, Io is convulsed by endless volcanic activity. Glowing a sulphurous yellow, its hellish landscape is dominated by seas of lava. Massive eruptions – the plume silhouetted on Io's edge is 140 kilometres (86 miles) high – are turning the moon inside out at a rate of one centimetre every 3,000 years.

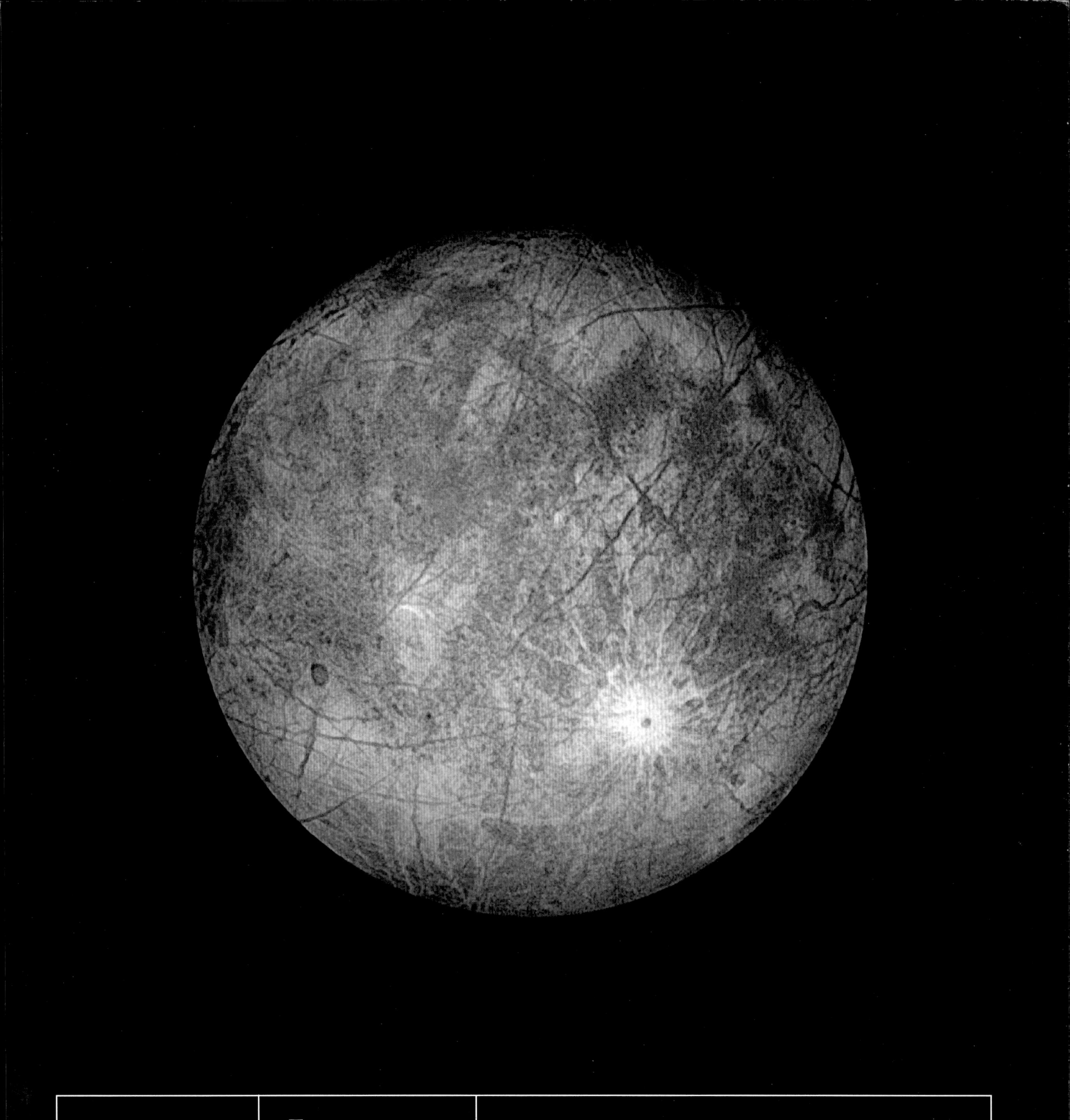

44.5 light minutes	*Europa* Moon	The same gravitational flux that has condemned Io to its infernal fate may have created a dark paradise 25 kilometres (15 miles) beneath the pack ice of Europa's surface. Here, warmed by thermal vents, a salty ocean may nurture life. If this is the case, we will know life can't be a fluke, that given the right conditions it is almost inevitable and we should expect to encounter it throughout the cosmos.

51 light minutes	*29P* Centuar	Centaurs are a species of comet found between the orbits of Jupiter and Neptune, believed to be fugitives from the first of the solar system's cometary reservoirs, the Kuiper Belt. With a nucleus 30 kilometres (20 miles) across, should this icy remnant of the primeval solar system fall into an orbit that approaches the Sun, it would form a spectacular comet, unrivalled in historic times.

83.4 light minutes	***Saturn*** Planet	Saturn is the second largest planet in our solar system, but it is also the most insubstantial. It is the only planet less dense than water – in the unlikely event of a large enough ocean being found, Saturn would float in it. Even without its rings, Saturn is an unusual shape: low density and fast rotation (a Saturnian day lasts just over 10 hours) combine to create flattened poles and a bulging equator.

83.4	*Saturn*	Locked in a procession 250,000 kilometres (150,000 miles) wide but under a kilometre (0.6 miles) thick, innumerable icy fragments cast a shadow over their captor. Shepherded into neat rows by the gravity of Saturn's 33 moons (Mimas stands guard here), the rings may be the remains of a single shattered satellite – if condensed they would form a sphere a mere 100 kilometres (62 miles) across.
light minutes	Rings	

83.4 light minutes	*Titan* Moon	The only moon with a substantial atmosphere, the clouds of Titan conceal an Earth-like landscape of riverbeds, dry lakes and erosion smoothed pebbles. But this familiar terrain is built from an alien chemistry: instead of liquid water, Titan has liquid methane; instead of rock, it has water ice; instead of soil, Titan has an oily mix of hydrocarbons. Even its volcanoes spew ice rather than molten rock.

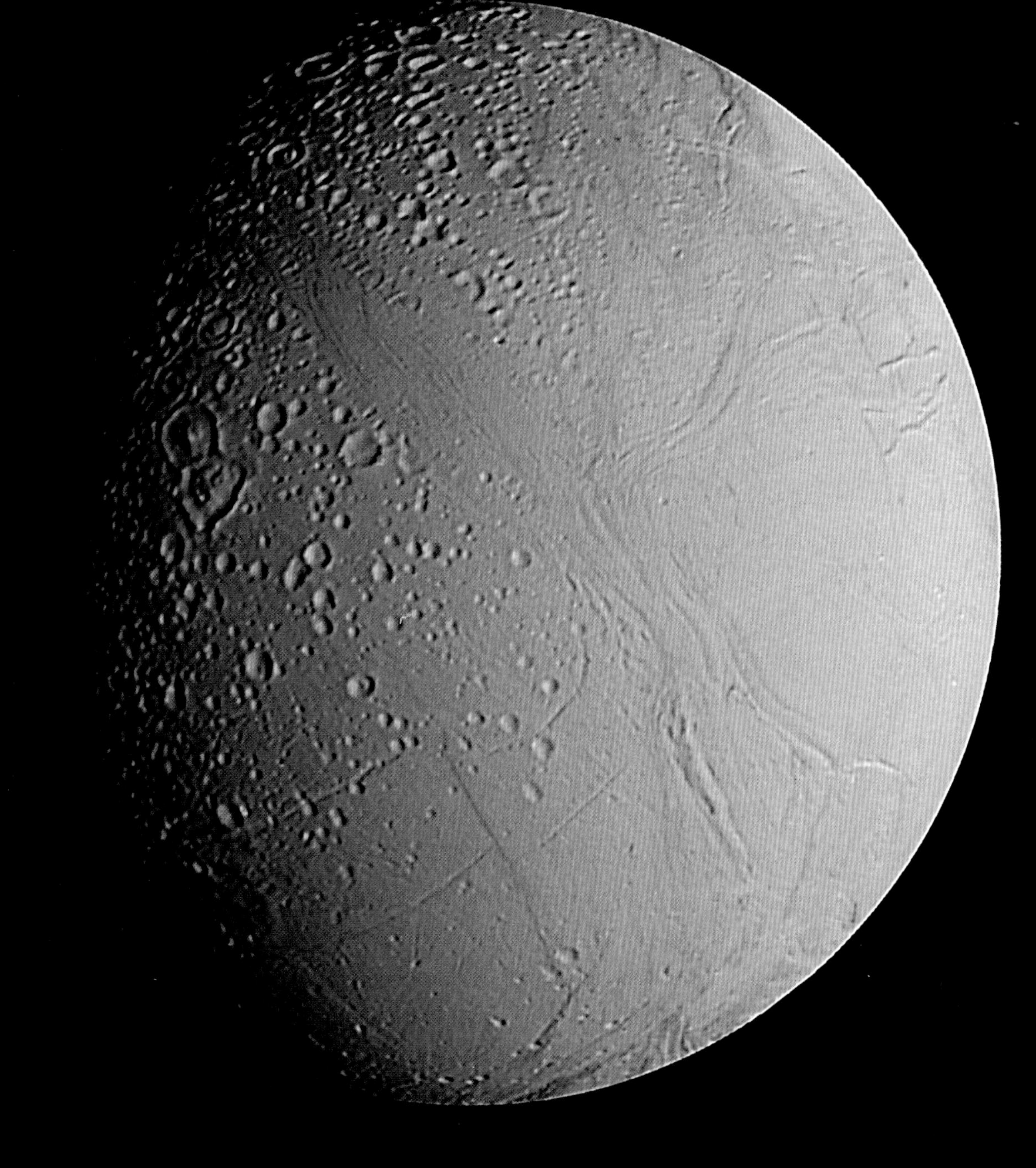

83.4 light minutes	*Enceladus* Moon	Snow-bright Enceladus reflects over 90 percent of the light that reaches it, making it the most brilliant body in the solar system. Kept fresh by a temperature of -201 °C (-330 °F), this moon's pistes are undoubtedly pristine – but the skiing here would disappoint: Enceladus' gravity is over 100 times weaker than Earth's.

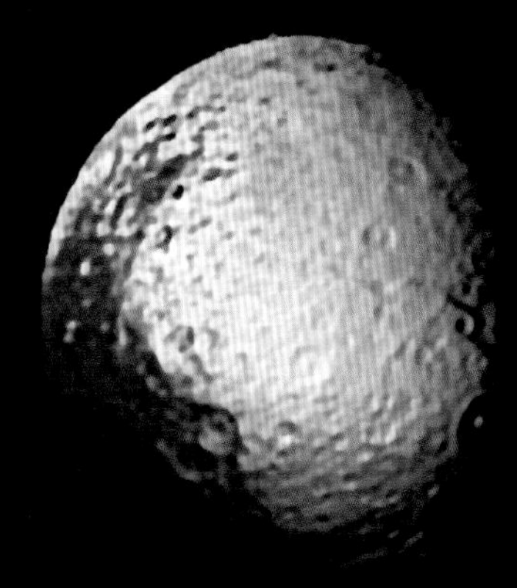

83.4	*Iapetus*	For half of its orbit Iapetus presents a bright, icy face; for the other half it reveals its dark side. One hemisphere of Saturn's third largest moon is coated with a sooty layer of one of the least reflective substances known. Some think that Iapetus was spray-painted with this dark material as it passed through a cloud of debris blasted off one of Saturn's outer moons by a meteorite impact.
light minutes	Moon	

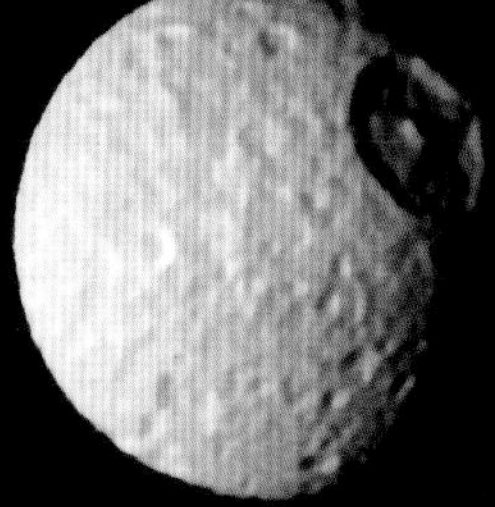

83.4

light minutes

Mimas

Moon

Who said Death Star? The spectacular Herschel crater covers nearly a quarter of Mimas' surface. Ten kilometres (six miles) deep, its outer walls rise five kilometres (three miles) high, while its central peak soars six kilometres (four miles) above the crater floor. Stress marks on the opposite hemisphere indicate that the impact came close to destroying the moon.

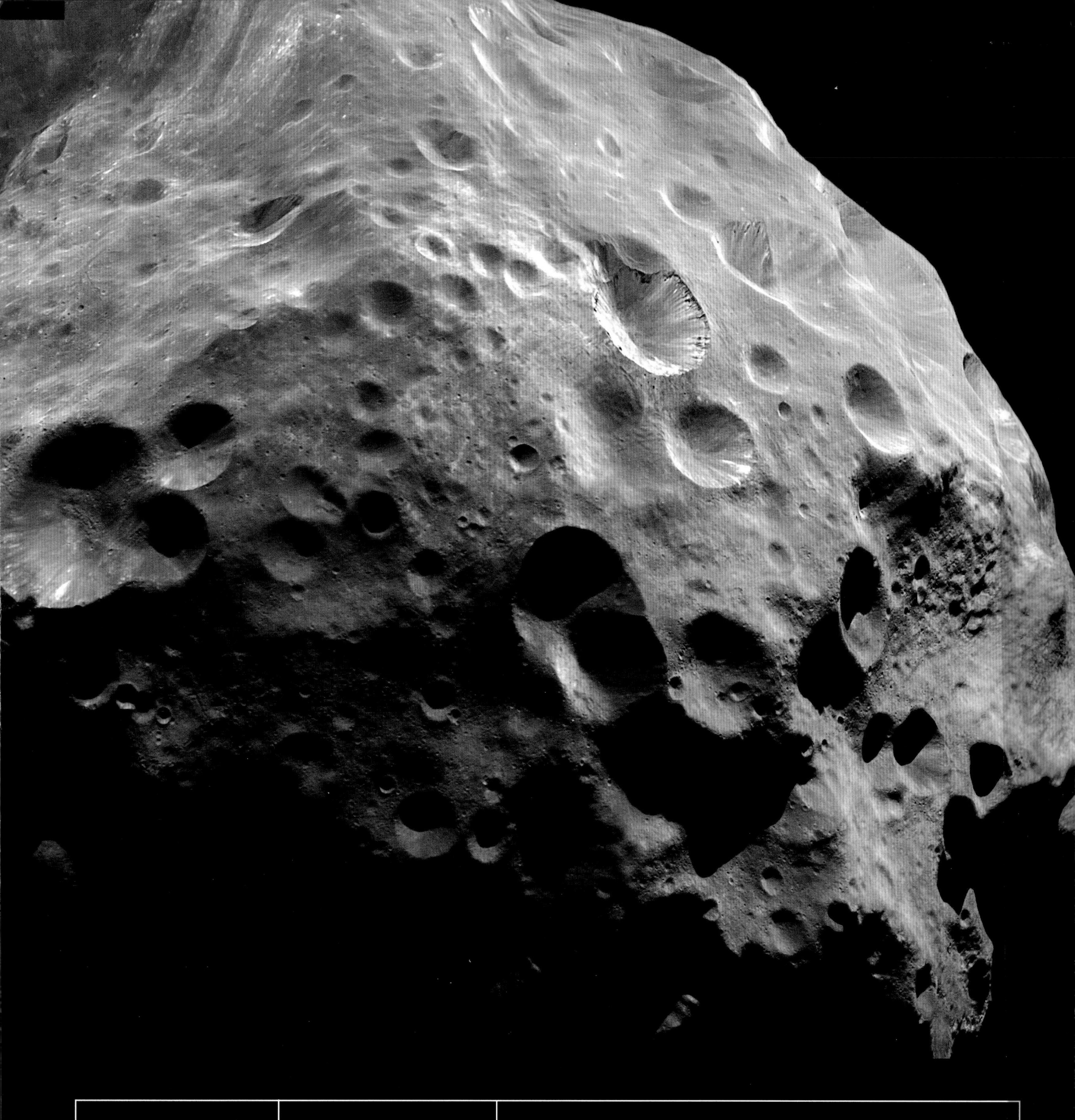

83.4 light minutes	*Phoebe* Moon	Only 214 kilometres (133 miles) across, Phoebe is not massive enough to pull herself into a perfect sphere. Her size, icy composition and unusual orbit mean she may be a relic of a bygone era, a captive member of a species of comet that has long since been banished to the fringes of the solar system. If this is so, we shall encounter Phoebe's long-lost sisters in the Kuiper Belt.

83.4 light minutes	*Tethys* Moon	Tethys is girdled by a vast fissure known as the Ithaca Chasma, seen here tracing the boundary between night and day. 100 kilometres (62 miles) wide and several kilometres deep, it covers two-thirds of Tethys' 3,330 kilometre (2,000 mile) circumference and is believed to have formed from the concussion of an impact that left a 400 kilometre (250 mile) crater on the other side of the moon.

163 light minutes	*Uranus* Planet	Uranus' northern hemisphere is emerging from the grip of a long, dark winter. Winters on Uranus are compounded by the fact that at the poles the Sun does not rise for 21 years as a consequence of the planet's extreme axial tilt. As sunlight returns, the frigid atmosphere warms stirring spring storms that blow bright clouds of crystallized methane before them at up to 420 kph (260 mph).

163

light minutes

Uranus

Rings

Uranus is encircled by a 10,000 kilometre (6,000 mile) belt of 11 dark and narrow rings. Uniquely, Uranus' rings are formed by a single-layered gravitational parade of soot-covered icy boulders with a diameter of no more than 30 metres (100 feet). The colour differences seen here, although exaggerated, are real, caused by differences in particle composition and size.

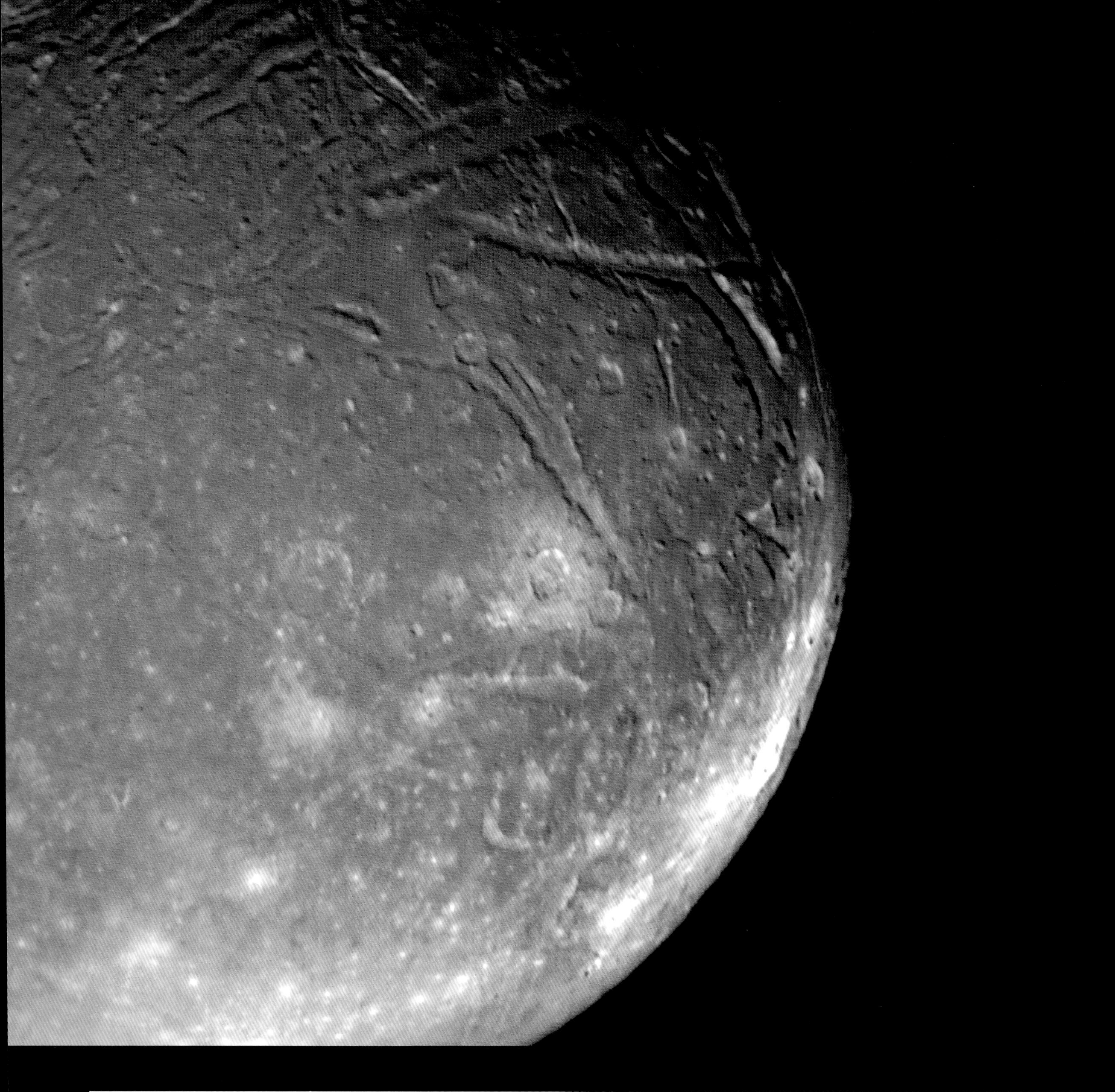

163 light minutes	*Ariel* Moon	Surrounding Uranus is a cast of 27 moons that take their names from the plays of William Shakespeare and the poems of Alexander Pope rather than from classical mythology. Ariel is a cold, geologically lifeless body of rock and ices now, but a maze of glacial flows and sunken valleys (chasmata) etched 10 kilometres (six miles) into its surface indicate that this hasn't always been the case.

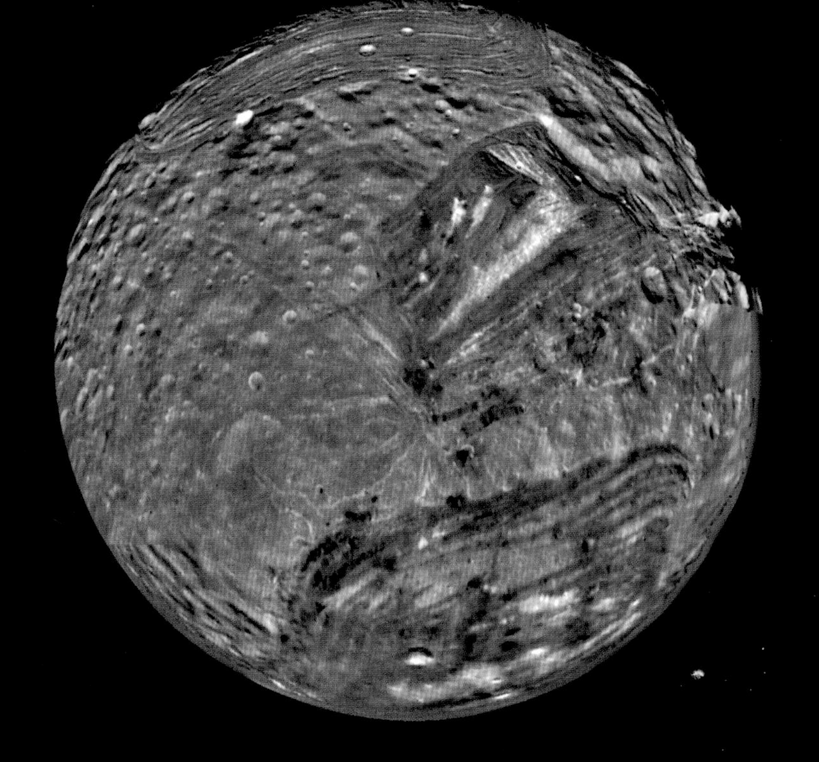

163 light minutes	*Miranda* Moon	A tempestuous past has left Miranda disfigured by a jumbled mass of geological scar tissue and gouged by concentric canyons up to 20 kilometres (12 miles) deep. It could just be that these are the marks of a moon stitched and re-stitched back together by gravity after up to five shattering impacts. A less dramatic explanation attributes the scars to periodic upwellings of metled ice.

163

light minutes

Oberon

Moon

Silhouetted against space on the upper rim of Oberon's heavily cratered face stands the peak of a six kilometre- (four mile-) high mountain. The moon has a diameter of 1,523 kilometres (974 miles) so proportionately this summit is higher than Olympus Mons on Mars, and would tower to over 50 kilometres (30 miles) if set on Earth.

163 light minutes	*Titania* Moon	With a diameter of 1578 kilometres (986 miles) Titania is the largest of Uranus' satellites. Like Ariel, she is covered with a web of chasmata. A slow freeze may explain this landscape: Titania may once have been hot enough to be liquid but as she cooled, her surface iced over first; then as her interior froze it expanded, cracking the already congealed crust and forming the fissures that we see today.

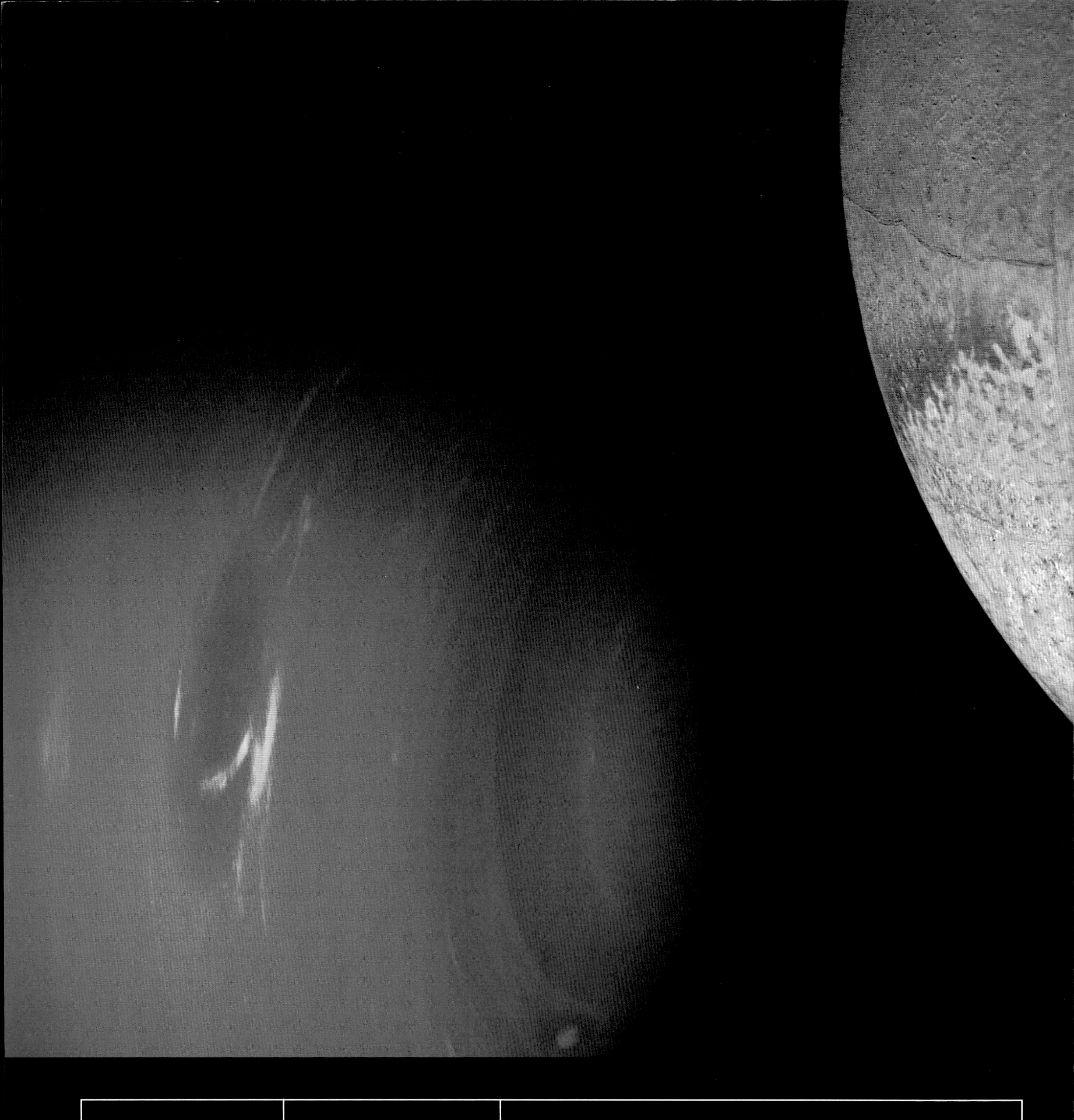

243

light minutes

Neptune

Planet

Beneath the 2,000 kph (1,200 mph) winds that lash Neptune's blue skies falls a hard rain of diamonds. Compressed from a boiling ocean of methane, the diamonds plummet thousands of kilometres towards the planet's core. The friction of their descent powers the storms seen on the surface and helps explain why Neptune radiates 2.6 times more energy than it receives from the Sun.

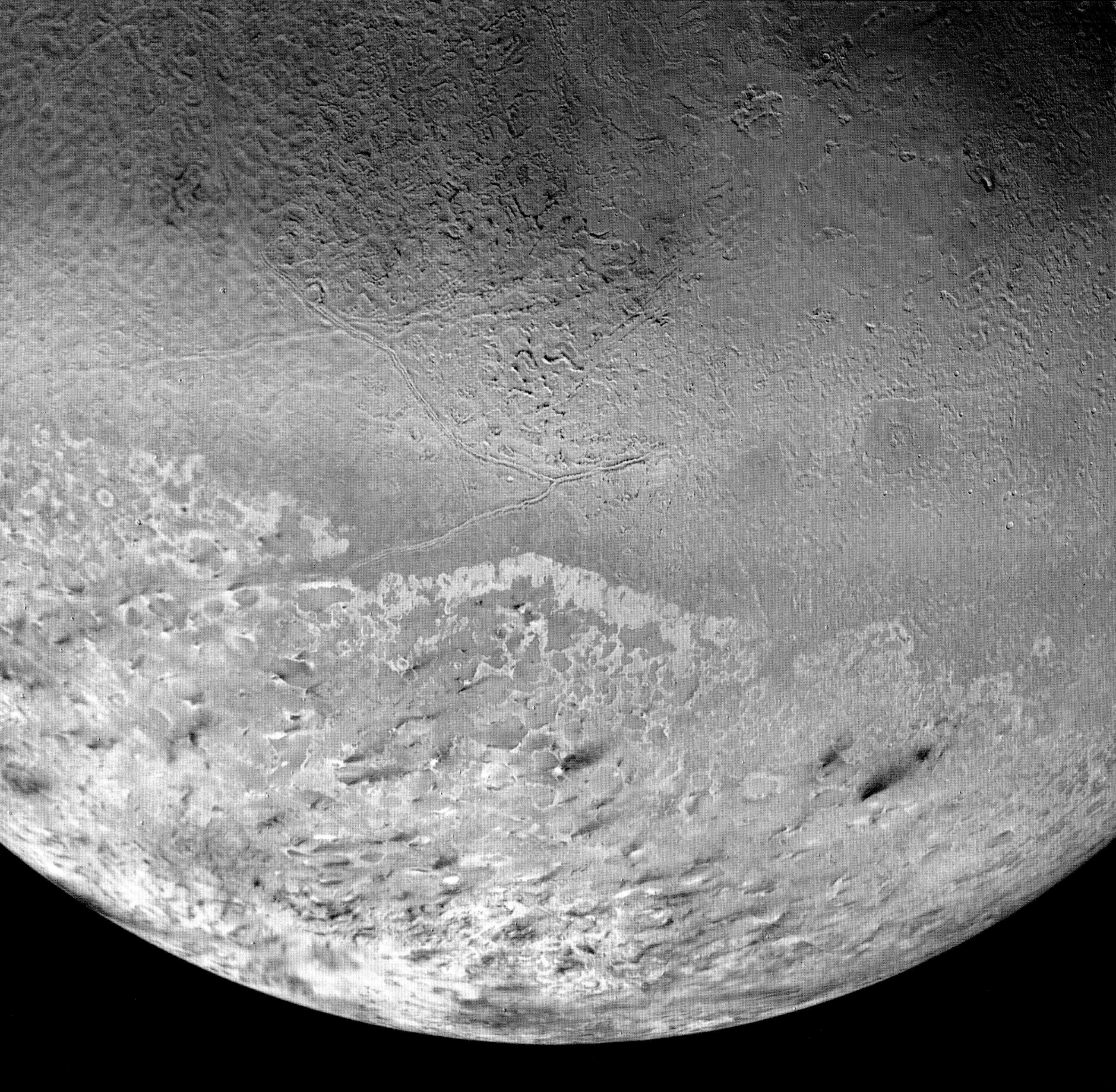

243 light minutes	*Triton* Moon	Triton was expected to be a cold, dead world. And it is cold, even colder than Pluto at -235 °C (-391 °F), but geologically speaking it is very much alive: erupting geysers throw dark plumes eight kilometres (five miles) into space. Alive for now, that is: a decaying orbit means Triton will collide with Neptune in the next ten million years, if it doesn't break up first.

249

light minutes

Pluto & Charon

Planet & moon

Locked in an elegant dance that sees them cross orbits, Neptune and Pluto share the burden of being the solar system's outer gatekeeper. But at the dark edges of the solar system not everything is as it seems: recent discoveries suggest Pluto and Charon are imposters – vagabond pretenders to the title of planet and moon hailing from that icy reservoir of short-period comets, the Kuiper Belt.

253 light minutes	*Halley's Comet* 1P/Halley Comet	Comet Halley is a familiar sight in the inner solar system, having visited Earth every 76 years since at least 240 BC, but its true habitat lies out here, in the Kuiper Belt, where it mixes with its less peripatetic ilk. This far from the Sun the comet is dormant, a honeycomb of dust and ice 15 kilometres (9 miles) long and half as wide. It will next return to light our night skies in 2062.

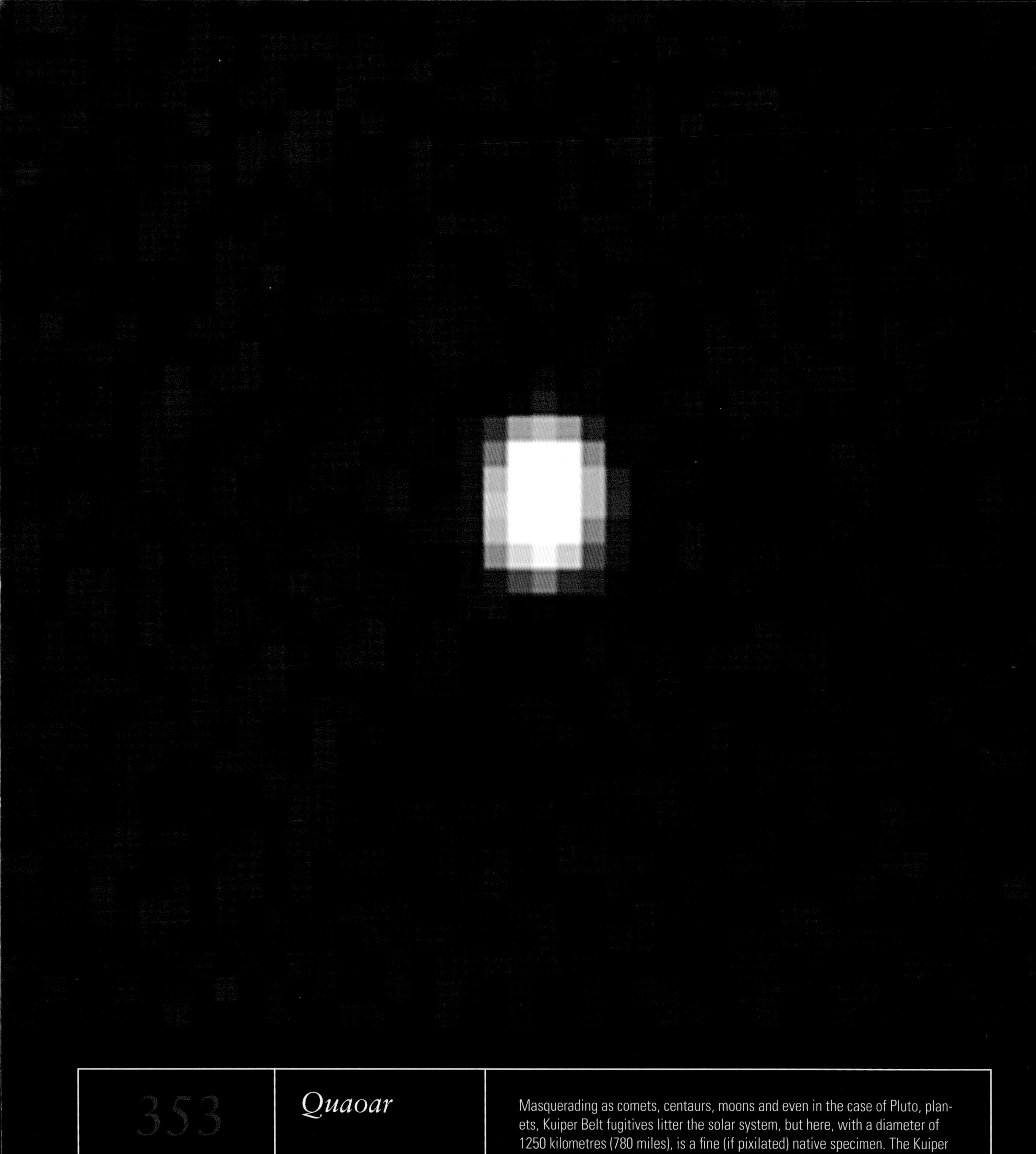

353 light minutes	*Quaoar* Kuiper Belt Object	Masquerading as comets, centaurs, moons and even in the case of Pluto, planets, Kuiper Belt fugitives litter the solar system, but here, with a diameter of 1250 kilometres (780 miles), is a fine (if pixilated) native specimen. The Kuiper Belt is a swarm of primordial icy bodies that originally inhabited the inner solar system but was banished to its fringes by the gravity of the gas giants.

745

light minutes

Sedna

Kuiper Belt Object

72 years away from the perihelion of its 11,487 year orbit, Sedna will approach to within 600 light minutes of Earth before receding again to nearly 8,000. With such an elliptical orbit Sedna may not be a Kuiper Belt Object, it could be an emissary from the even more distant Oort Cloud. Two-thirds the size of Pluto, Sedna is the coldest object in the solar system at -240 °C (-400 °F).

797 light minutes	*Voyager 1* Spacecraft	Voyager 1 is the most distant man-made object in the universe. Travelling at 0.006 percent of the speed of light (61,000 kph or 38,000 mph), it is also the fastest. Launched in 1977 on a Grand Tour of Jupiter and Saturn, Voyager 1 is now approaching the heliopause, a turbulent region of space where the solar wind begins to falter before it finally gives way to the interstellar medium.

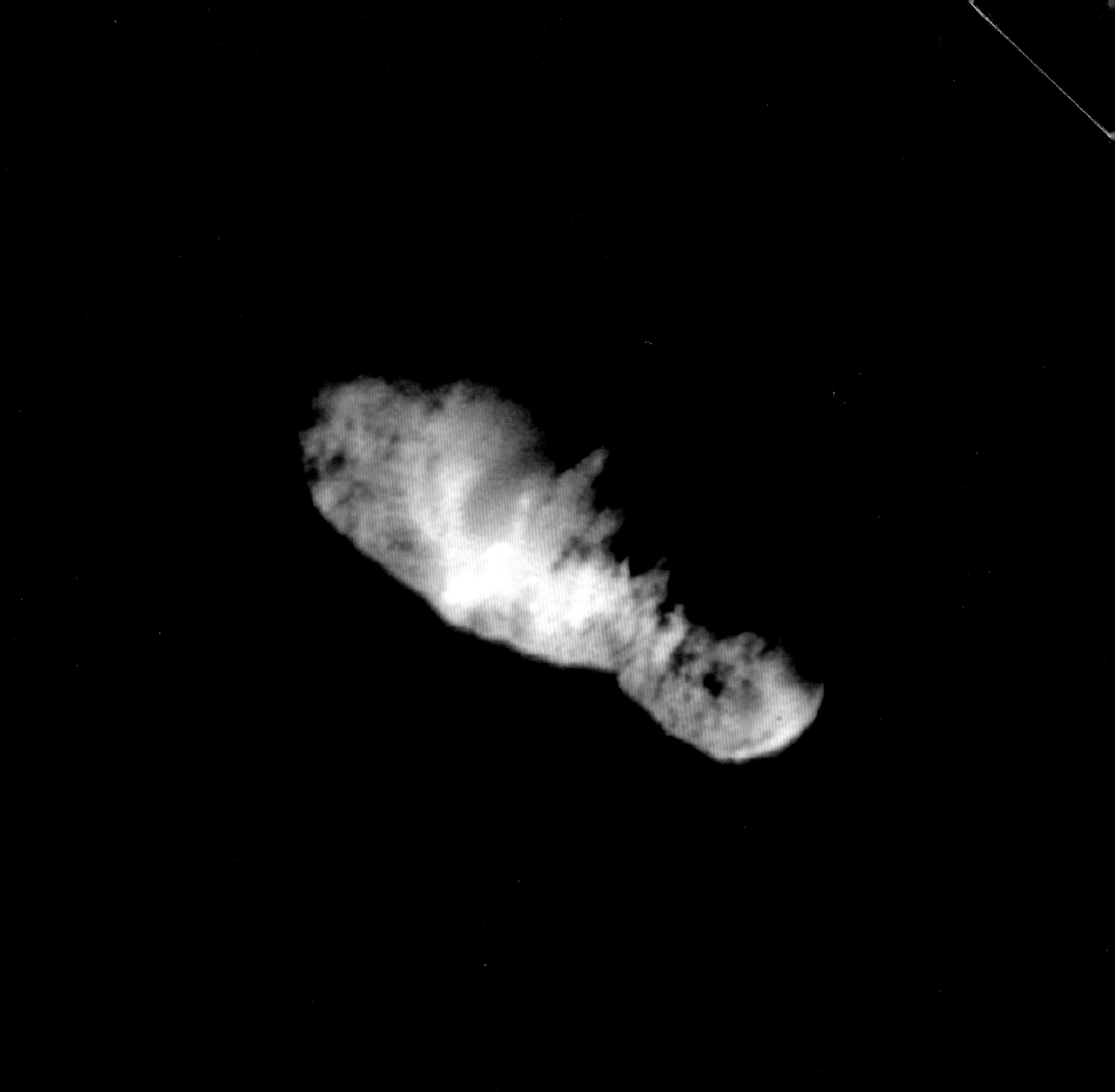

1 light year	*Oort Cloud* Oort Cloud Object	At the very edge of the solar system – so far away that the Sun is no longer the brightest star in the sky – lies the Oort Cloud. The icy exiles here are occasionally nudged out of their distant orbits to plunge towards the Sun as long-period comets. The Oort Cloud is so remote we can only infer its existence, so here Comet Borrelly doubles as an Oort Cloud Object.

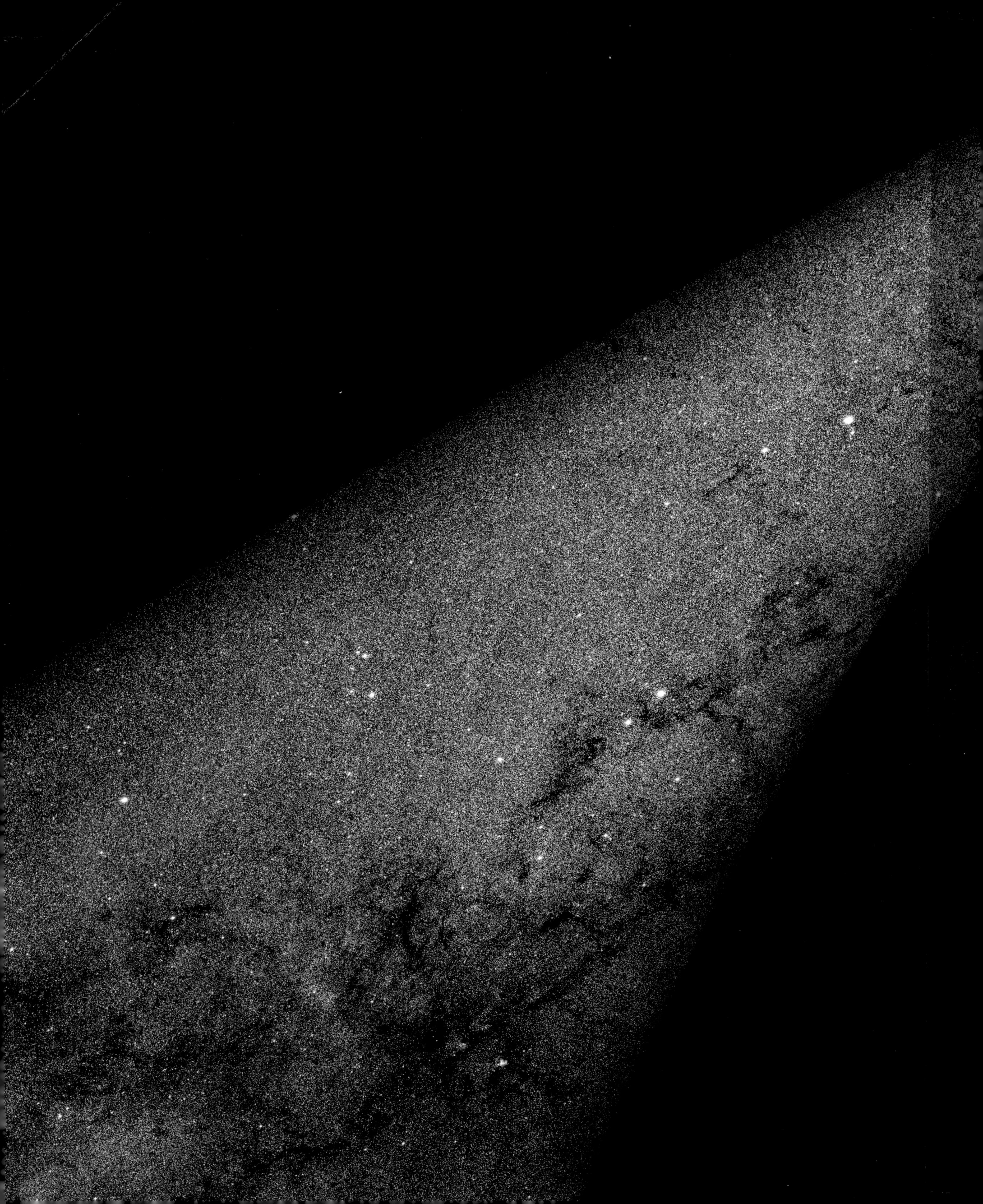

2 our galaxy

4.2 light years	*Proxima Centauri* Red dwarf star	While only eight light minutes to our Sun, it will take you four years to reach the next one. Proxima Centauri might only be a tenth of the size of our Sun and 18,000 times dimmer, but as this X-ray photograph reveals, it can punch above its weight. Like many red dwarfs, Proxima Centauri is a flare star – explosive releases of energy can triple the star's brightness in a matter of minutes.

180 light years	*RX J 185635-3754* Neutron star	The three grainy smudges lined out left of centre may not immediately impress, but they represent the progress of a fugitive stellar corpse as big as Manhattan Island but 55 trillion times denser, travelling at 389,000 kph (240,000 mph). Forged at the heart of a supernova one million years ago, RX J 185635-3754 is the closest known neutron star. And it's heading our way.

440

light years

Pleiades

NGC 1432

Open star cluster

The ancient Greeks recognized the Pleiades, or Seven Sisters, as the children of Atlas (the Titan condemned to hold up the heavens) and Pleione, daughter of Oceanus. The sisters are (clockwise, from left) Alcyone, Asterope, Taygeta, Celaeno, Electra, Merope and in the centre, Maia. The Seven Sisters are actually eight as Asterope is a double star. Their blue veil is a passing dust cloud.

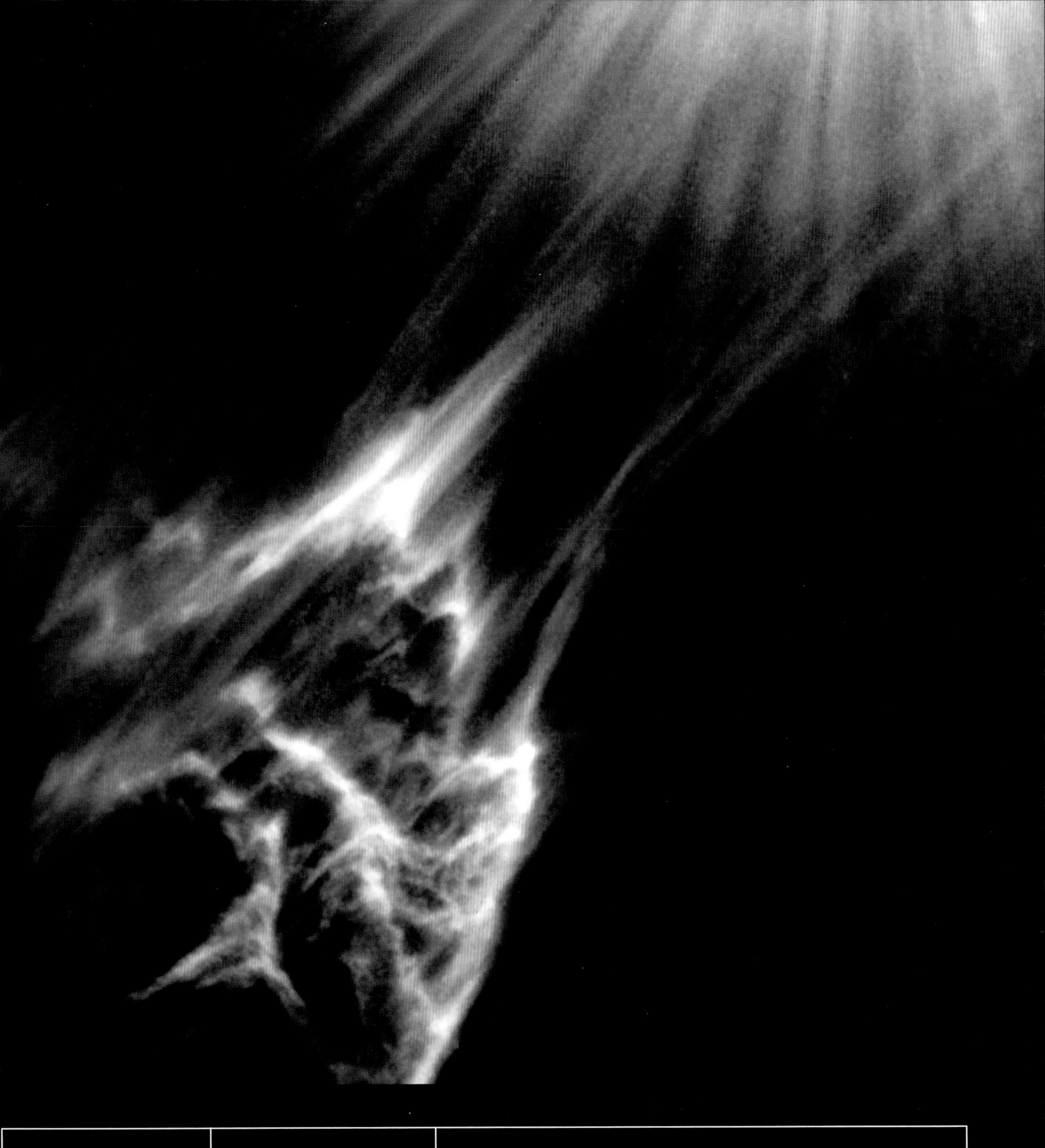

440	*IC 349*	Pleiad Merope's searchlight rays illuminate this cold, dark cloud of interstellar dust and gas. Passing within a galactic hair's breadth of the star – a mere 500 billion kilometres (300 billion miles) – the cloud is being torn apart by its encounter. Radiation pressure is sifting the smaller dust particles from the cloud like chaff from wheat, creating the bright streaks seen above.
light years	Reflection nebula	

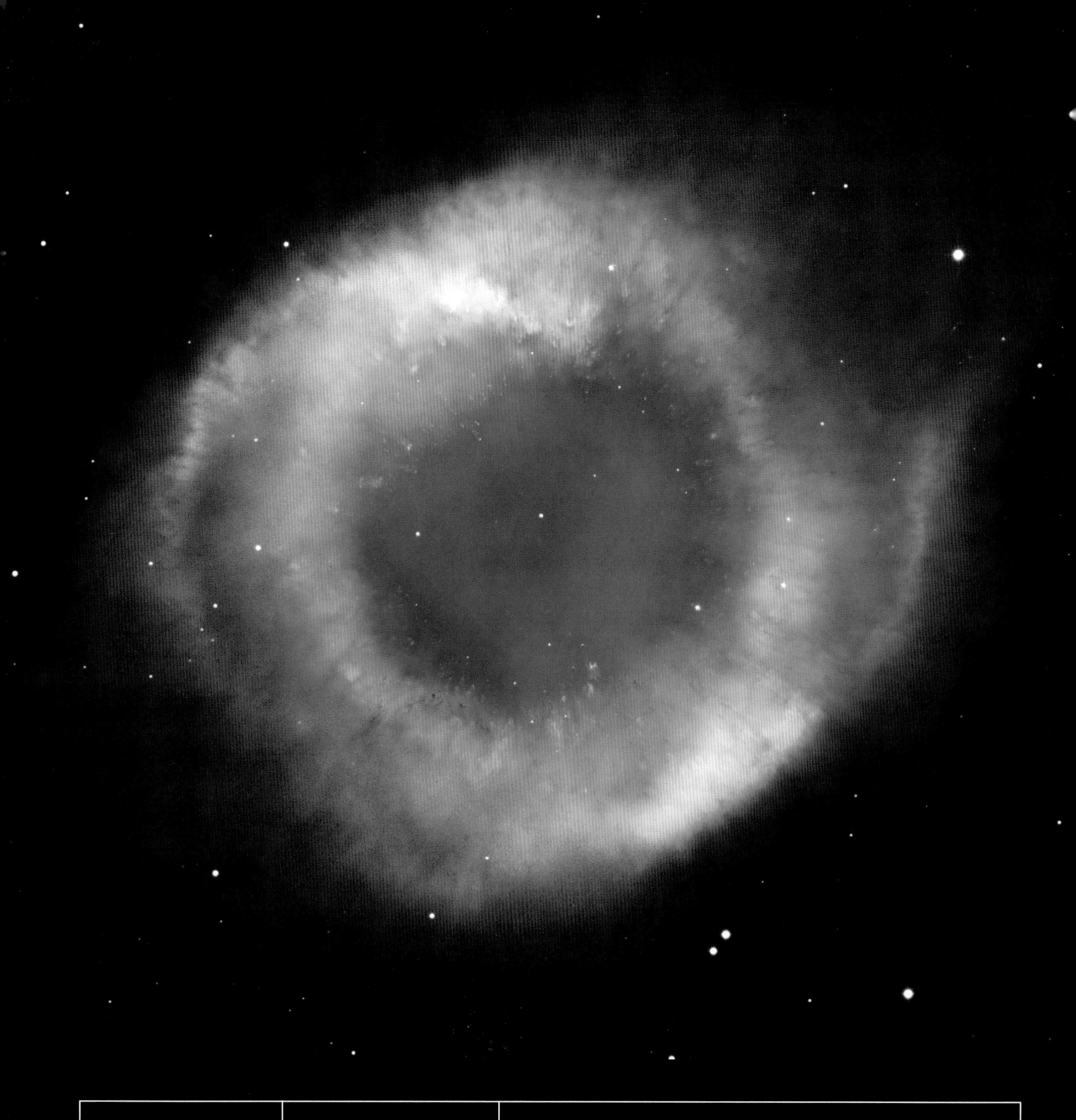

450

light years

Helix Nebula

NGC 7293

Planetary nebula

Pointing straight at us, this trillion-mile barrel of fluorescing gas is the final flowering of a Sun-like star. Planetary nebulae are short-lived phenomena lasting no more than 50,000 years, blossoming as red giant stars cast off their outer mantles and dwindle to white dwarfs. In five billion years there will be no need to travel 450 light years to see such a sight: we'll have one at home.

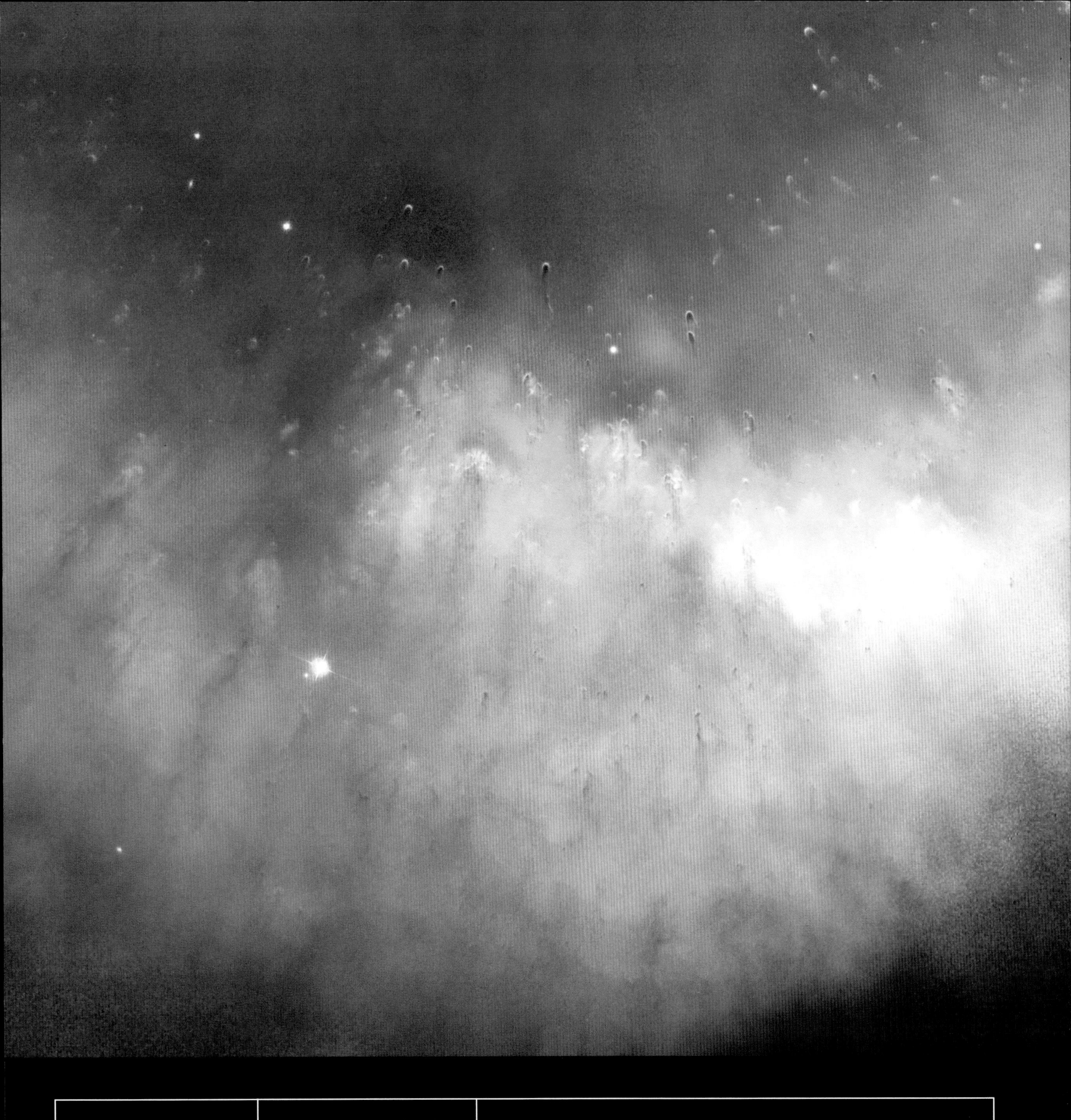

450 light years	*Helix Nebula* NGC 7293 {detail} Planetary nebula	A forest of thousands of tentacle-like filaments, embedded along the inner rim of the nebula, reach back toward their central star, a small, super-hot white dwarf. These filaments form as cold globules of previously ejected dust and gas are buffeted by the hotter stellar wind streaming from the doomed star. This wind is blowing the nebula before it at 100,000 kph (62,000 mph).

500

light years

Barnard 68

Dark nebula

This looming hole in space is actually a molecular cloud. At -263 °C (-440 °F) it is as cold as it is dark – and one of the coldest objects in the universe. Yet with a little compression, these frigid clouds are the unlikely birthplace of new stars. At half a light year wide, if our solar system was engulfed by such a cloud we would be consigned to a night sky utterly devoid of stars.

500

light years

Chamaeleon Complex

Reflection nebula

A heavenly wash of dust and starlight creates these watercolour skies. Thin shells of carbon dust (literally the ashes of dead stars) bask in the light of two bright young suns, scattering blue light to form the central reflection nebula, while surrounding molecular clouds laced with thicker dust block the light from more distant stars.

815

light years

Pencil Nebula

NGC 2736

Supernova remnant

These clouds mark the leading edge of a 500,000 kph (300,000mph) tsunami that has been ploughing through space for 11,000 years. Unleashed by the explosive death of a massive star, or supernova, this shockwave has slowed from its original velocity of 35 million kph (22 million mph), but the violence of its passage is still enough to light up the interstellar medium as it passes.

1 thousand light years	*HH 32* Herbig-Haro Object	Still nurtured by the molecular cloud surrounding it, this newborn star ejects some of the in-falling material in a 1.1 million kph (700,000 mph) hiccup that launches perpendicularly opposed jets 2 trillion kilometres (1.2 trillion miles) into space. A hallmark of the final stages of stellar construction, these distinctive celestial blowtorches are known as Herbig-Haro Objects.

thousand
light years

Dumbbell Nebula M27

Planetary nebula

Discovered by Charles Messier on July 12th 1764, the nature of the Dumbbell Nebula intrigued early observers. In 1785 William Herschel dubbed such objects 'planetary' because their round, disc-like shape reminded him of his recent discovery, the planet Uranus and the name has stuck. He guessed they signified the construction of solar systems, rather than their destruction.

1.5 **thousand** light years	*Orion Nebula* M42 Emission nebula	We now enter the great Orion Molecular Cloud. Stretching over 100 light years, it invisibly occupies most of the Orion constellation. But where this dark nebula has spawned new stars, they illuminate the chill clouds from which they coalesced, giving rise to some of the local universe's most remarkable sights, including this huge stellar nursery at the centre of Orion's sword.

1.5 **thousand** light years	*Orion Nebula* M42 {detail} Emission nebula	2.5 light years across, this cavern at the heart of the Orion Nebula has opened to reveal the infernal details of stellar construction. Over 700 suns are maturing here, harvesting material from the cloud, waiting for their chance to shine. Many are robed in swirling discs of dust – embryonic solar systems – assuring us that planet formation is common throughout the universe.

1.5

thousand
light years

Trapezium

M42 {detail}

Open star cluster

The Trapezium, the Orion Nebula's four hottest, most massive stars, flared into life one million years ago, lighting up the nebula's previously dark clouds with a blaze of high-energy radiation. They are surrounded by a swarm of brown dwarfs, dim stars too small to ignite the fusion of hydrogen in their cores, shining only as gravitational energy is converted into heat.

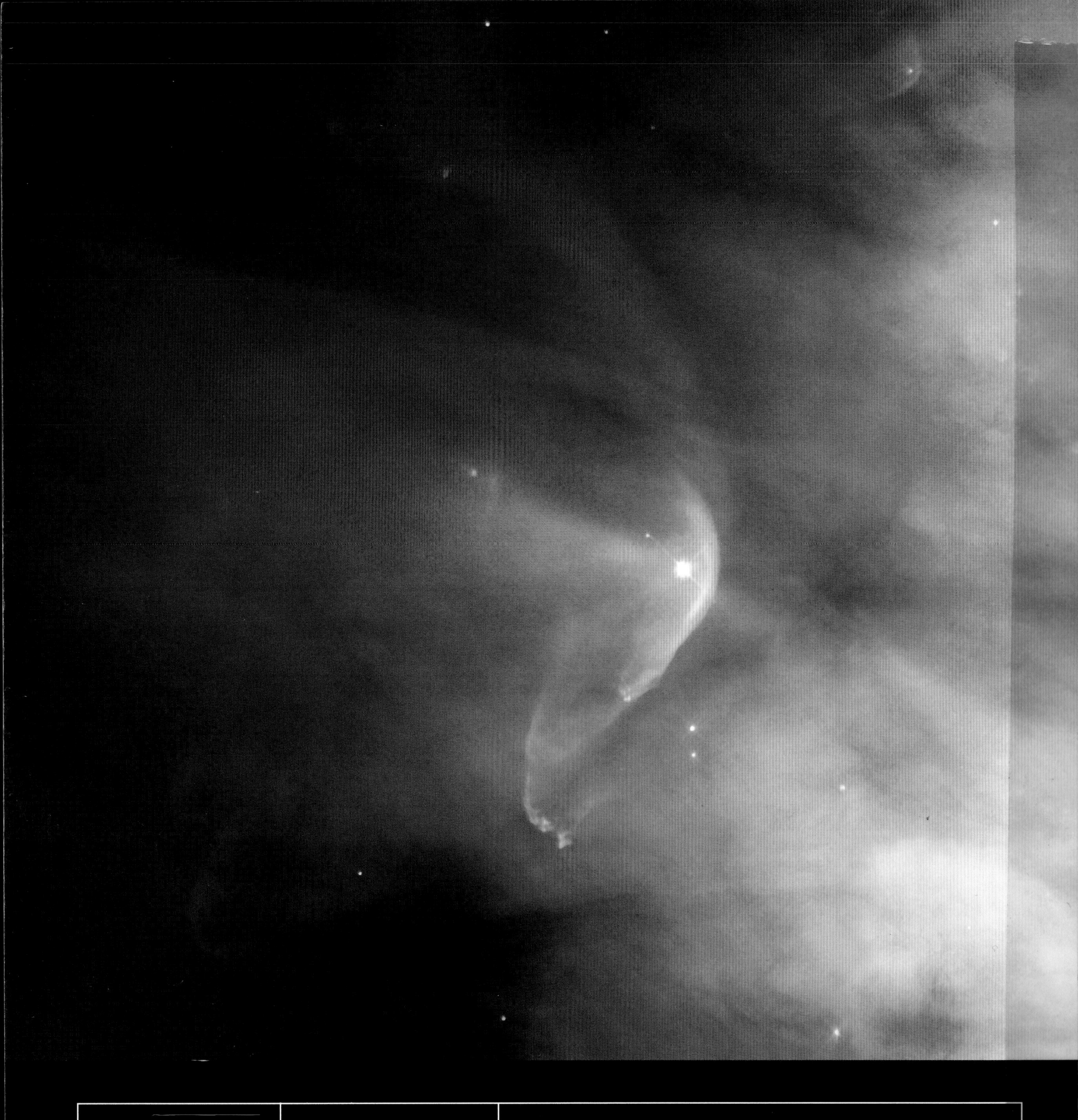

1.5

thousand
light years

LL Ori

M42 {detail}

T-Tauri star

A stellar nursery is a turbulent environment. Shock waves rock the cradle as infant stars begin to wake. As protostars grow, the stream of hot particles they emit intensifies, creating a gusting wind that tears into, and ultimately dismantles, the cocoon of dust and gas that once nurtured them. Such kindling stars are classified as T-Tauri.

1.5

thousand
light years

Flame Nebula

NGC 2024

Emission nebula

The Flame Nebula is another beacon illuminating the greater darkness of the Orion Molecular Cloud. To our eyes, its hydrogen clouds glow in the searing light of Alnitak, one of the stars of Orion's belt, but infrared inspection reveals its belly to be heavy with stars. Pregnant for the last million years, one day the Flame Nebula may upstage its more brazen neighbours.

1.5

thousand
light years

NGC 1999

Dark & reflection nebula

From the darkness there shall come light. This ominous keyhole is actually a dense condensation of dust and gas blocking out all light from the star behind it. Such clouds are known as Bok globules and, locked within, protostars are thought to be quickening. Behind, the all encompassing Orion Molecular Cloud glows like fog around a flashlight in the beams of V380 Orionis.

1.5 thousand light years	*Horsehead Nebula* IC 434 Dark & emission nebula	Perhaps the most famous denizen of the Orion Molecular Cloud, this dark knight rears a full light year above a glowing sea of hydrogen. Sigma Orionis lights this celestial shadowplay from offstage: hydrogen clouds fluoresce in the glare of the star's ultraviolet radiation while a dense pillar of dust in the foreground presents a striking equine silhouette.

1.5

thousand
light years

HH 34

Herbig-Haro Object

At the base of this heavenly waterfall lies a Herbig-Haro Object in machine gun-, rather than flamethrower-, mode. The juvenile hooligan is spitting material into space not as a funnelled jet but in distinct bursts. It is thought such expulsions may assist star formation by carrying away excess angular momentum that might otherwise prevent in-falling material from reaching the star.

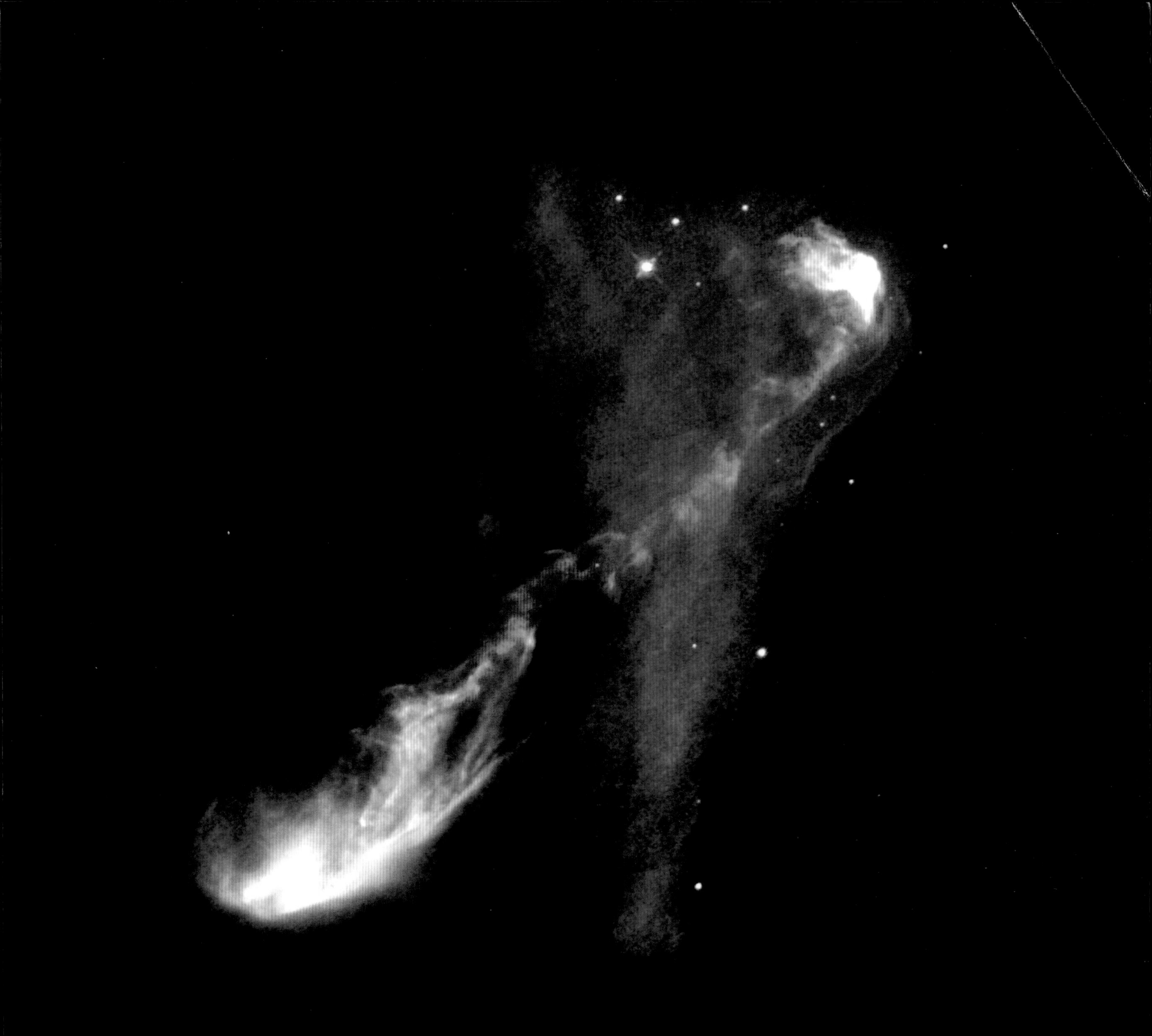

1.5 **thousand** light years	*HH 47* Herbig-Haro Object	Leaving the constellation of Orion, we find a more normally behaved Herbig-Haro Object on the edge of the Gum Nebula. Our protostar has yet to emerge from its cocoon (bottom left) but a five trillion kilometre (three trillion mile) plume of hot gas betrays its existence. The plume is twisted which may mean our stellar infant has been disturbed by another star, possibly an unseen sibling.

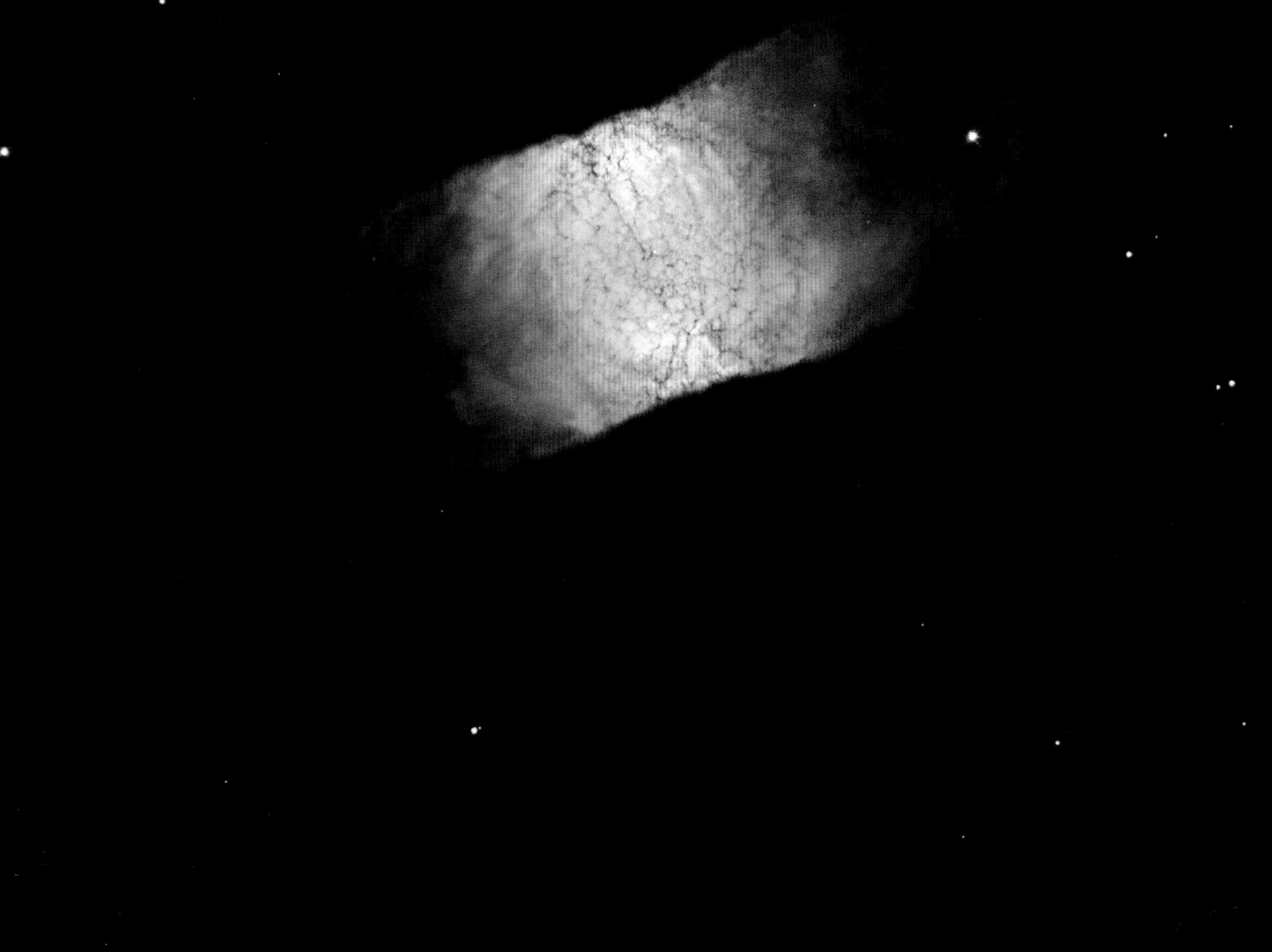

1.9

thousand
light years

Retina Nebula

IC 4406

Planetary nebula

Seen side on, the Retina Nebula is a dust-speckled torus of hydrogen, oxygen and nitrogen heated by the glare of its internal star until it fluoresces. Despite launching scalding blooms of gas (typically 12,000 °C or 20,000 °F) across vast areas of space at 700,000 kph (440,000 mph), a planetary nebula is a mild-mannered stellar catastrophe when compared with a supernova.

thousand
light years

Eight-burst Nebula NGC 3132

Planetary nebula

An interloper basks at the centre of this limpid pool. The faint companion, not the bright star, has conjured this panorama out of thin air – extremely thin air. With a typical density around 1000 particles per cubic centimetre, planetary nebulae are about a septillion (a trillion trillion) times less substantial than the air that we breathe.

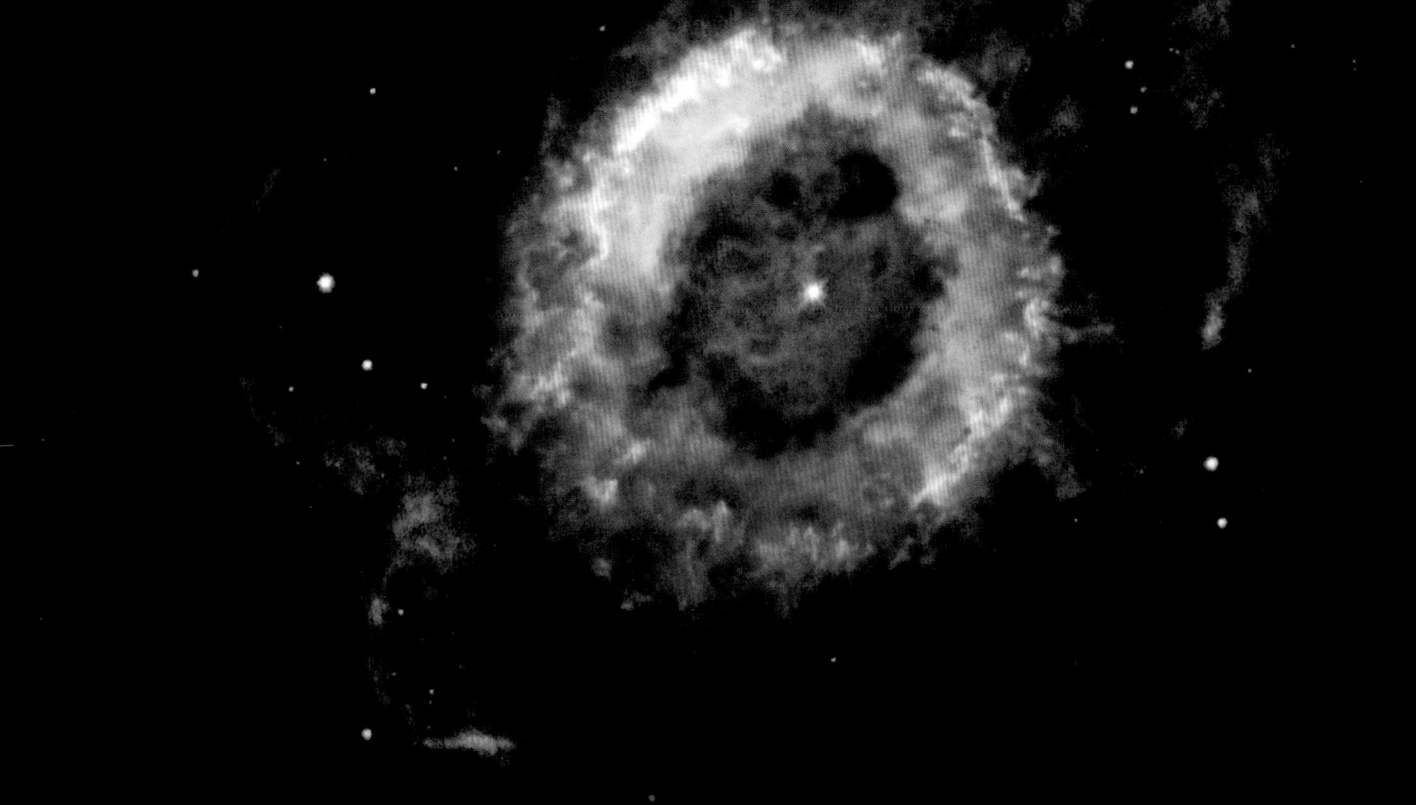

2 **thousand** light years	*Little Ghost Nebula* NGC 6369 Planetary nebula	Despite the pyrotechnics that accompany their births, there can be little doubt stars save their greatest performance for the final act. Over 1,500 planetary nebulae have been catalogued but the galaxy contains many thousands more – 95 percent of all stars are destined to step off the main sequence and slide into retirement as planetary nebulae.

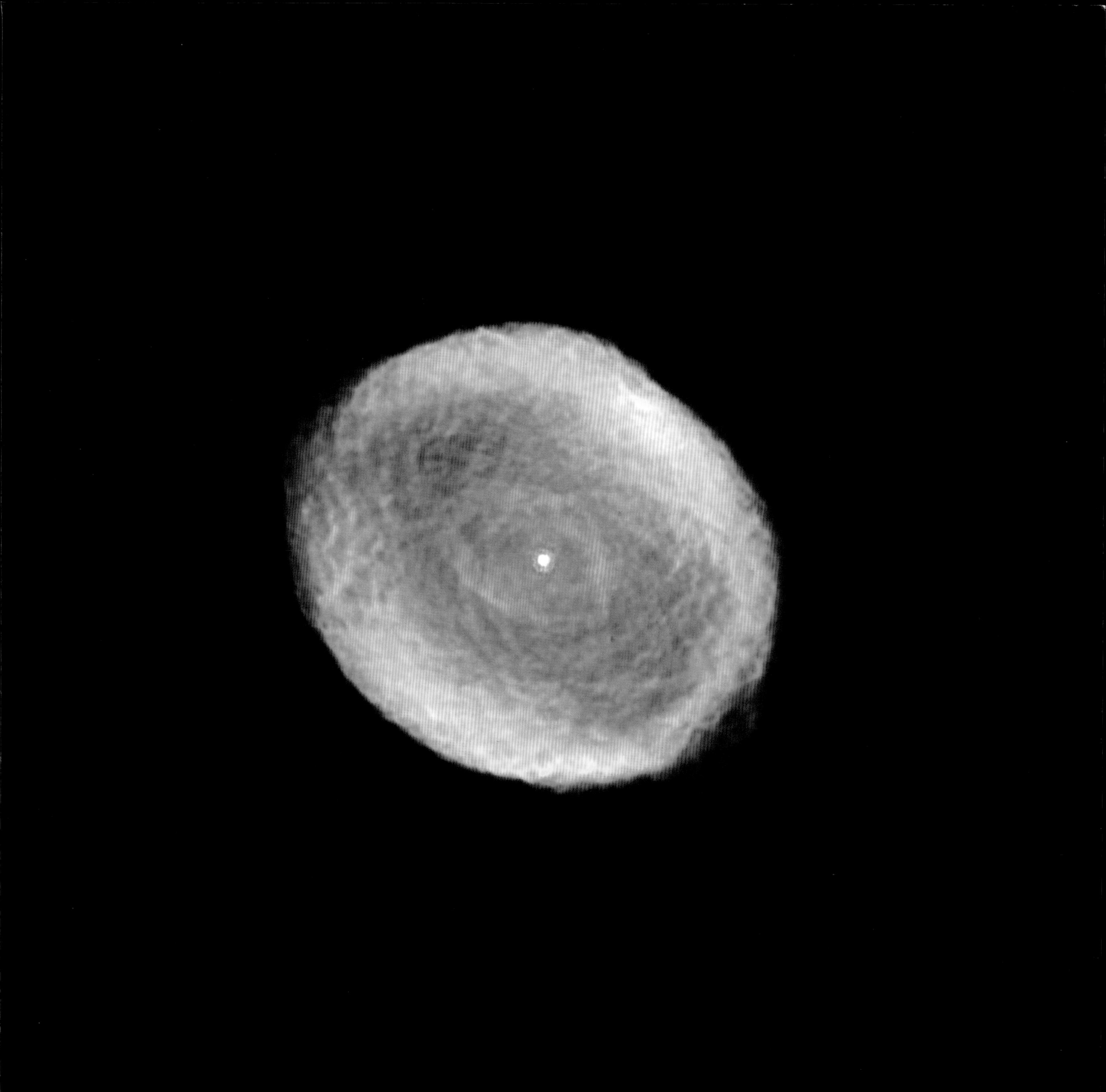

2 thousand light years	*Spirograph Nebula* IC 418 Planetary nebula	A tenth of a light year across, this youthful nebula owes its jewel-like structure to the unpredictable fluctuations of its central star. A mere tenth of a light year? Some perspective is necessary here: that is enough room to engulf our solar system 100 times over. And what happens to solar systems caught inside such nebulae? No-one is entirely certain, but the prognosis isn't good.

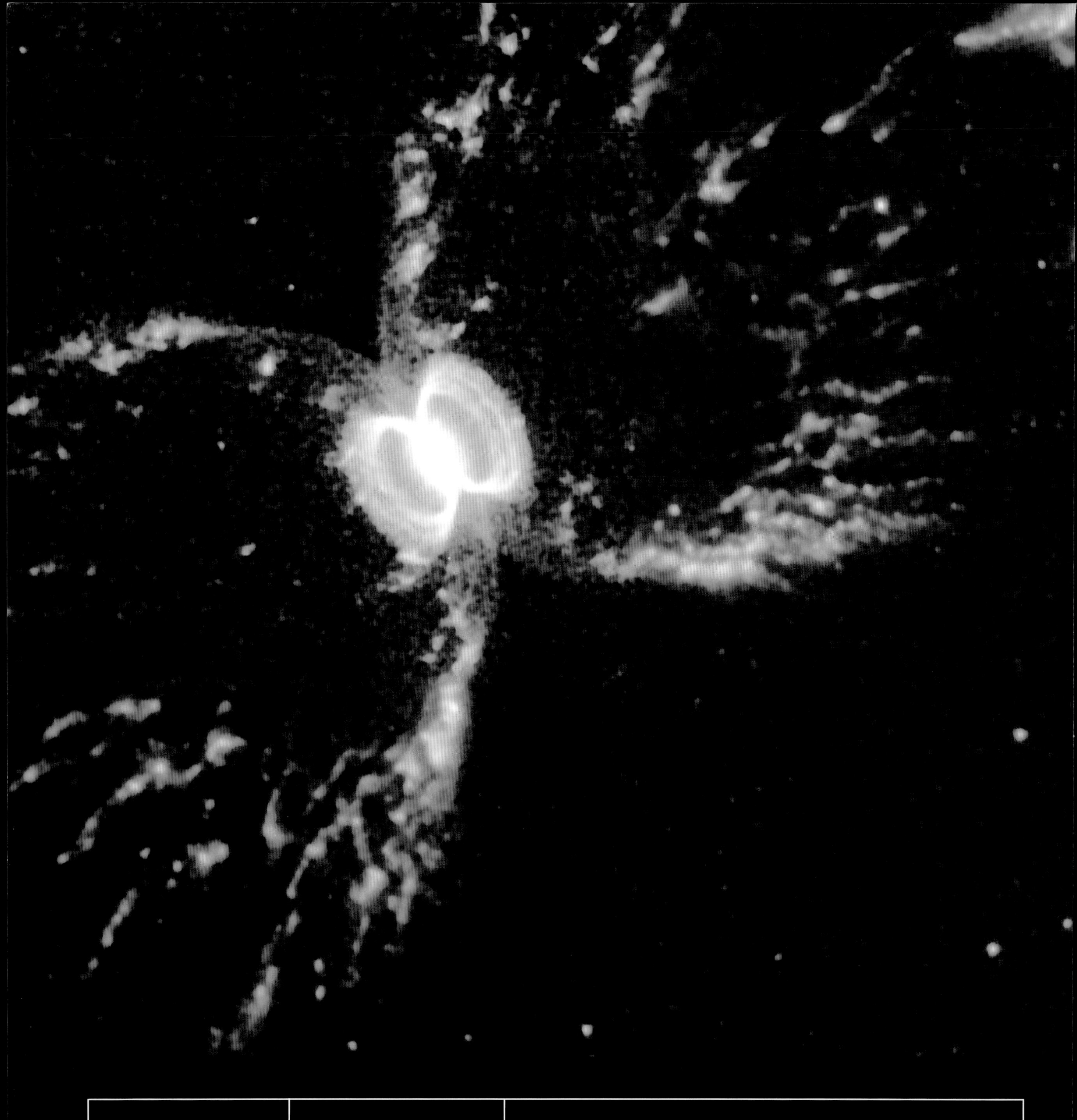

2 **thousand** light years	*Southern Crab Nebula* He2-104 Planetary nebula	The bright heart of these nested hourglasses masks the macabre feast of a stellar ghoul. A white dwarf is devouring the swollen outer layers of its red giant partner. Sucking down more than it can digest our ghoul has violently ejected the surfeit of material on more than one occasion, creating these interlocking luminescent splatters of nitrogen.

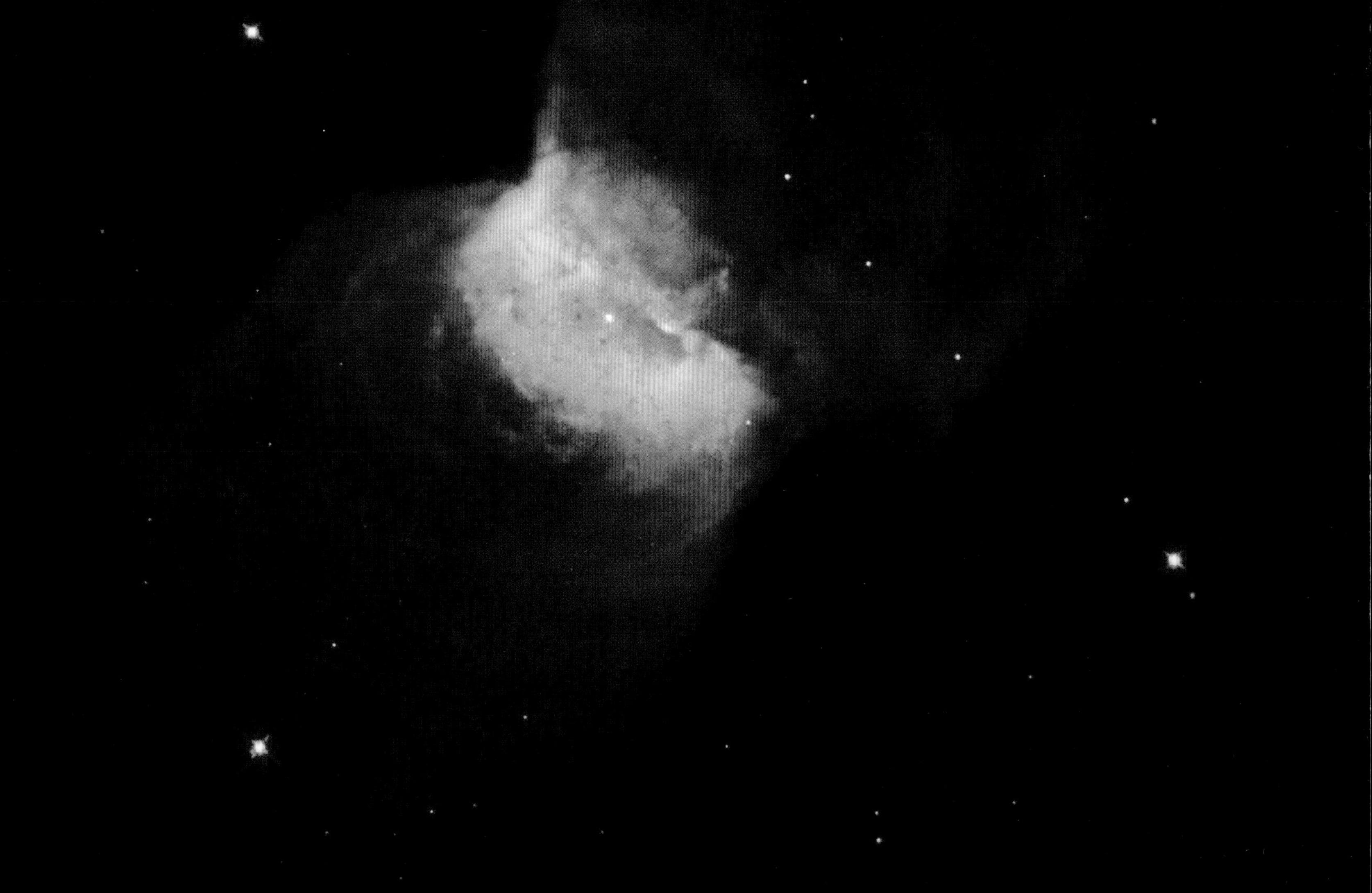

2 **thousand** light years	*NGC 2346* Planetary nebula	This butterfly, or bipolar, nebula conceals another incidence of stellar cannibalism. This time though, it's the red giant that swallowed its neighbour. Enveloped by the swollen body of the giant as it expanded, the companion star spiralled inward, expelling a disc of gas. This disc is now being inflated into two distinct lobes by ultraviolet radiation from the giant's remnant.

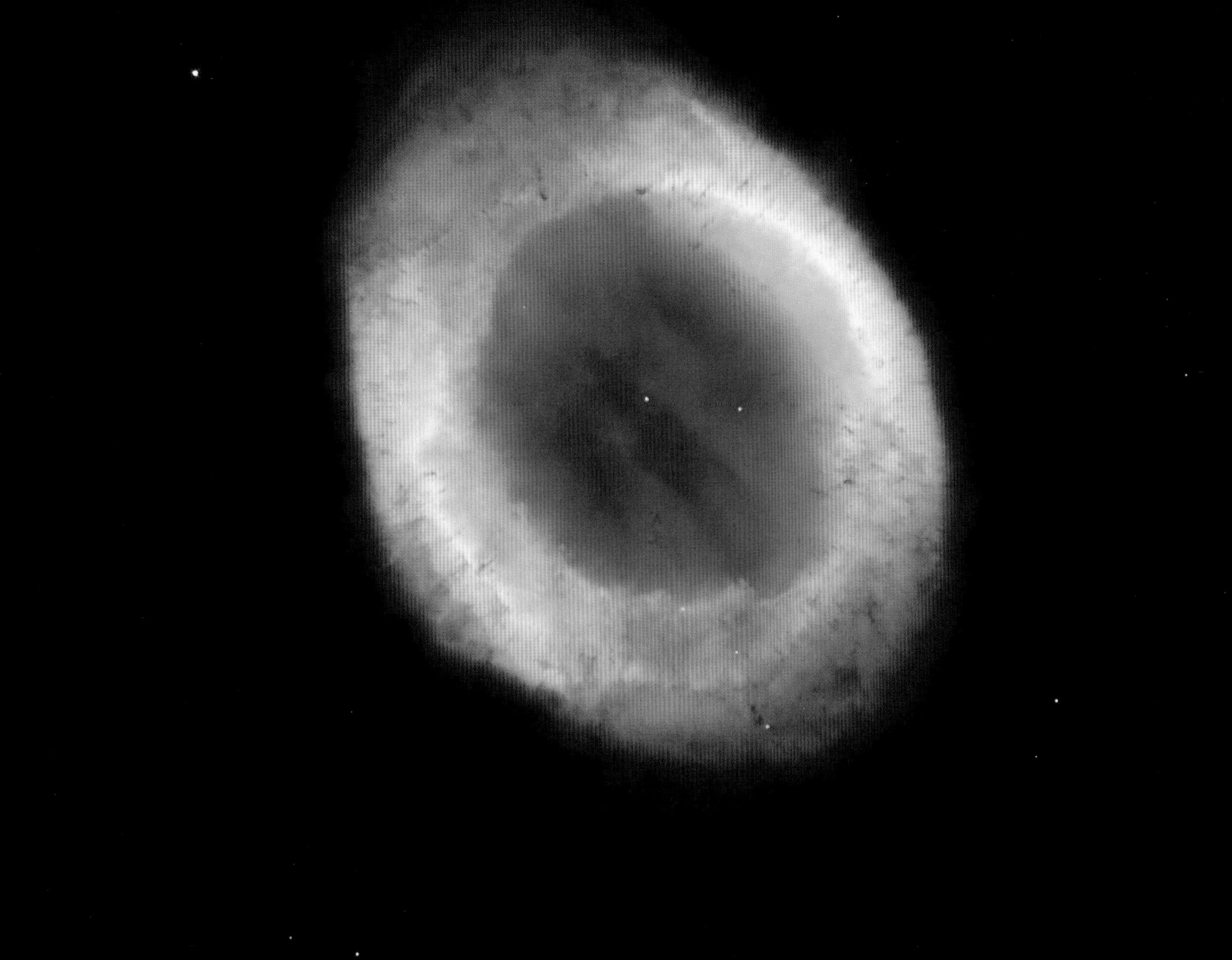

2 thousand light years	*Ring Nebula* M57 Planetary nebula	Like the Helix Nebula, we stare straight down the barrel of the famous Ring Nebula. Such smoking guns play an important role in the galactic ecosystem. As planetary nebulae expand they enrich the interstellar medium with the fruits of nuclear fusion – elements such as carbon, nitrogen and oxygen – and sow the seeds for new generations of stars, planets and people.

thousand
light years

Twin Jet Nebula M2-9

Planetary nebula

The snowflakes of the stellar realm, no two planetary nebulae are the same, but as we have seen, most (but not all) come in two distinct flavours: donut and butterfly. This young specimen, about 1,200 years old, is another butterfly nebula thought to be typical of binary star systems. Approximately 10 percent of planetary nebulae display such a bipolar formation.

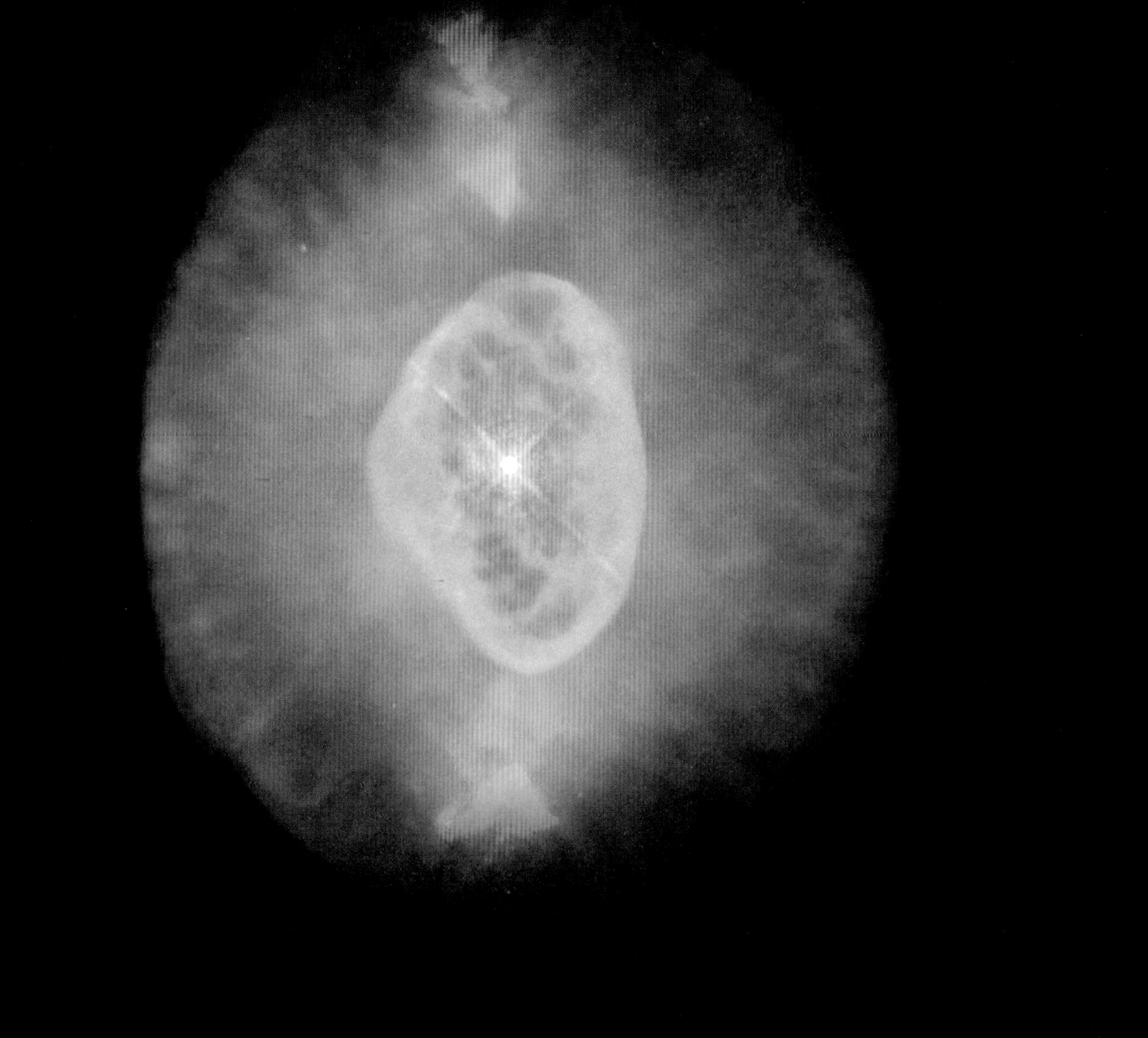

2.2 **thousand** light years	*Blinking Eye Nebula* NGC 6826 Planetary nebula	Planetary nebulae blossom only for a few millionths of a star's life span – in the blink of a galactic eye this nebula will be but a memory. Having ceased fusion, white dwarfs are no more than stellar embers radiating the last of their energy into space. Cooling rapidly, within a handful of millennia they can no longer ionize their gas cloud and the nebula fades back into the dark.

2.3

thousand
light years

Red Rectangle

HD 44179

Proto-planetary nebula

Our census of planetary nebulae continues with this immature specimen. The Red Rectangle is not strictly a planetary nebula as its central star isn't yet hot enough to ionize the material it is shedding. Instead, this nebula shines with reflected light. Complex chains of hydrocarbons have been detected here, reminding us that we too were originally forged in a stellar furnace.

2.4 **thousand** light years	*Elephant's Trunk* IC 1396 Dark nebula	Flaming in the stellar wind of an off-stage massive star, the Elephant's Trunk is a dark globule of dust and gas threading its way across emission nebula IC 1396. But seen in infrared, it is pierced by the heat of star formation. Embryonic stars glow red as they condense, heating up as they pull more material from the cloud. When they reach 10 million °C (18 million °F) they will ignite.

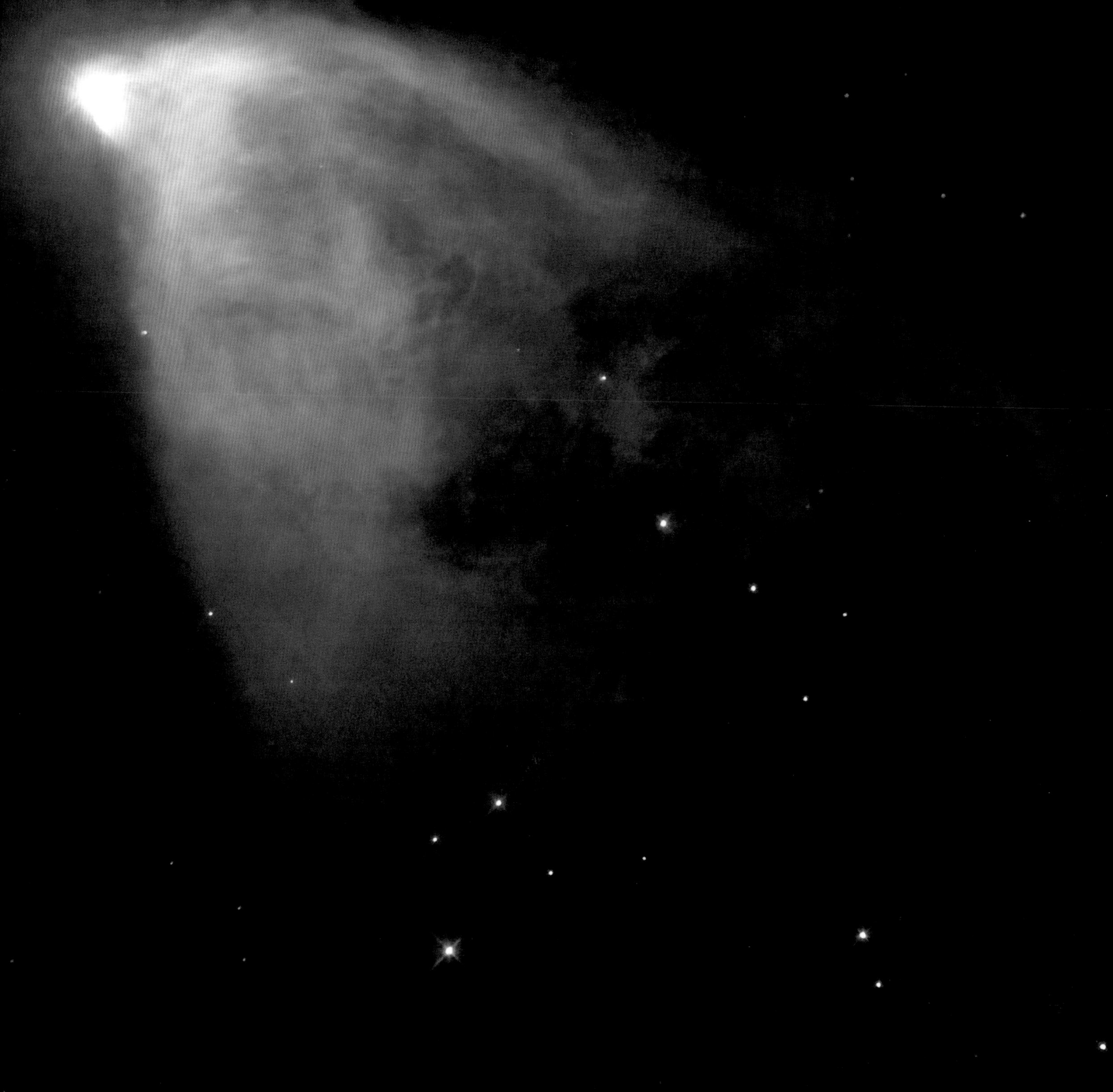

2.5

thousand
light years

Hubble's Variable Nebula

NGC 2261

Dense knots of gas sliding across the face of R Monocerotis cast moving shadows into this nebula, changing its appearance week by week. It is named, like the telescope that produced this image, after the astronomer Edwin Hubble. In the 1920s, Hubble realised that many 'nebulae' were in fact distant galaxies. From this deduction he was able to infer the true extent of the universe.

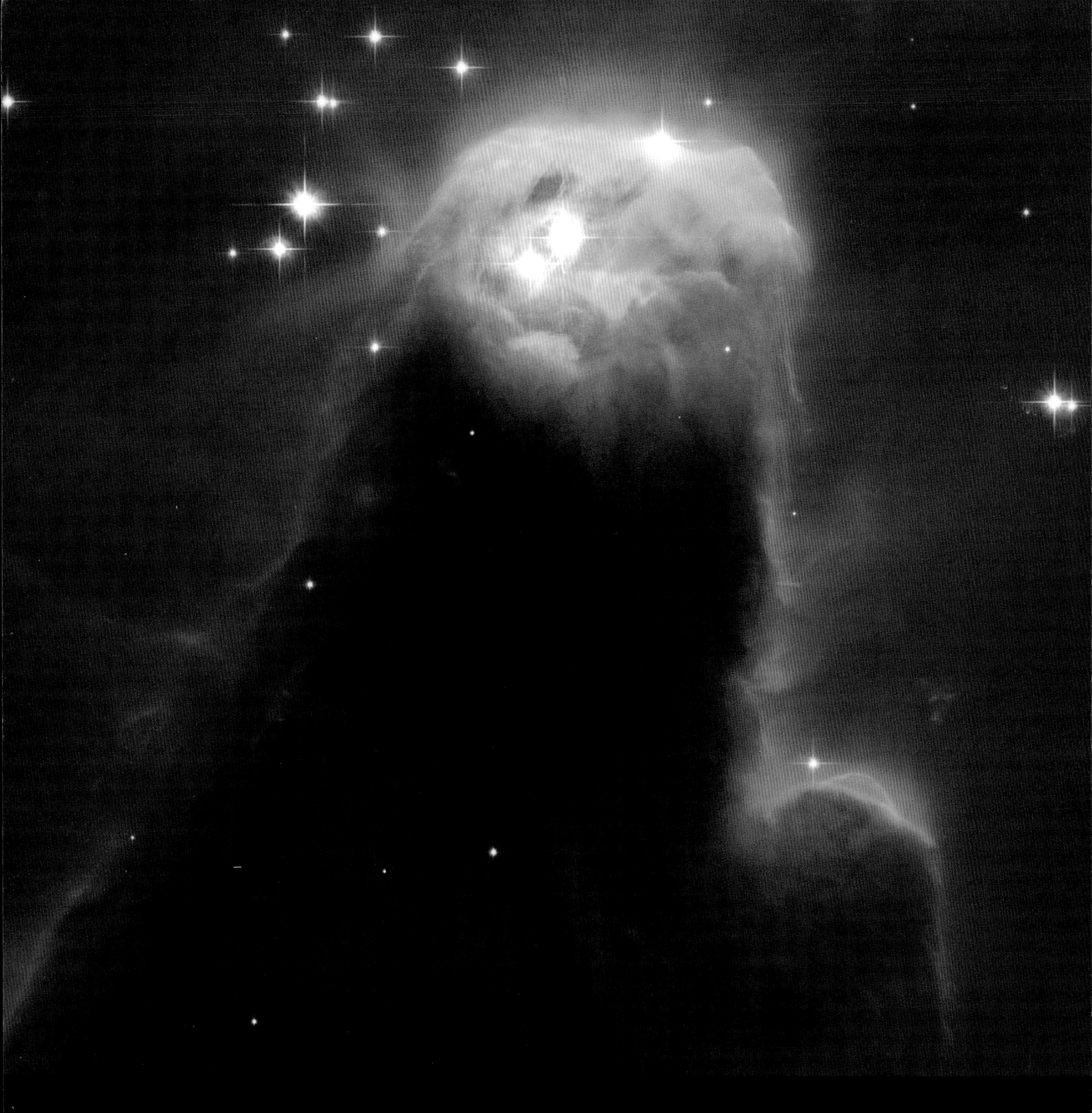

2.5 **thousand** light years	*Cone Nebula* NGC 2264 Dark and emission nebula	A leviathan towering seven light years into a blood-red sky, the Cone Nebula is of the same ilk as the Horsehead Nebula: a dense pillar of dust and gas backlit by fluorescing gas. Intense radiation from nearby stars is eroding this monstrous apparition, boiling hydrogen out of its clouds. But while it stands, cocooned within, protostars are forming.

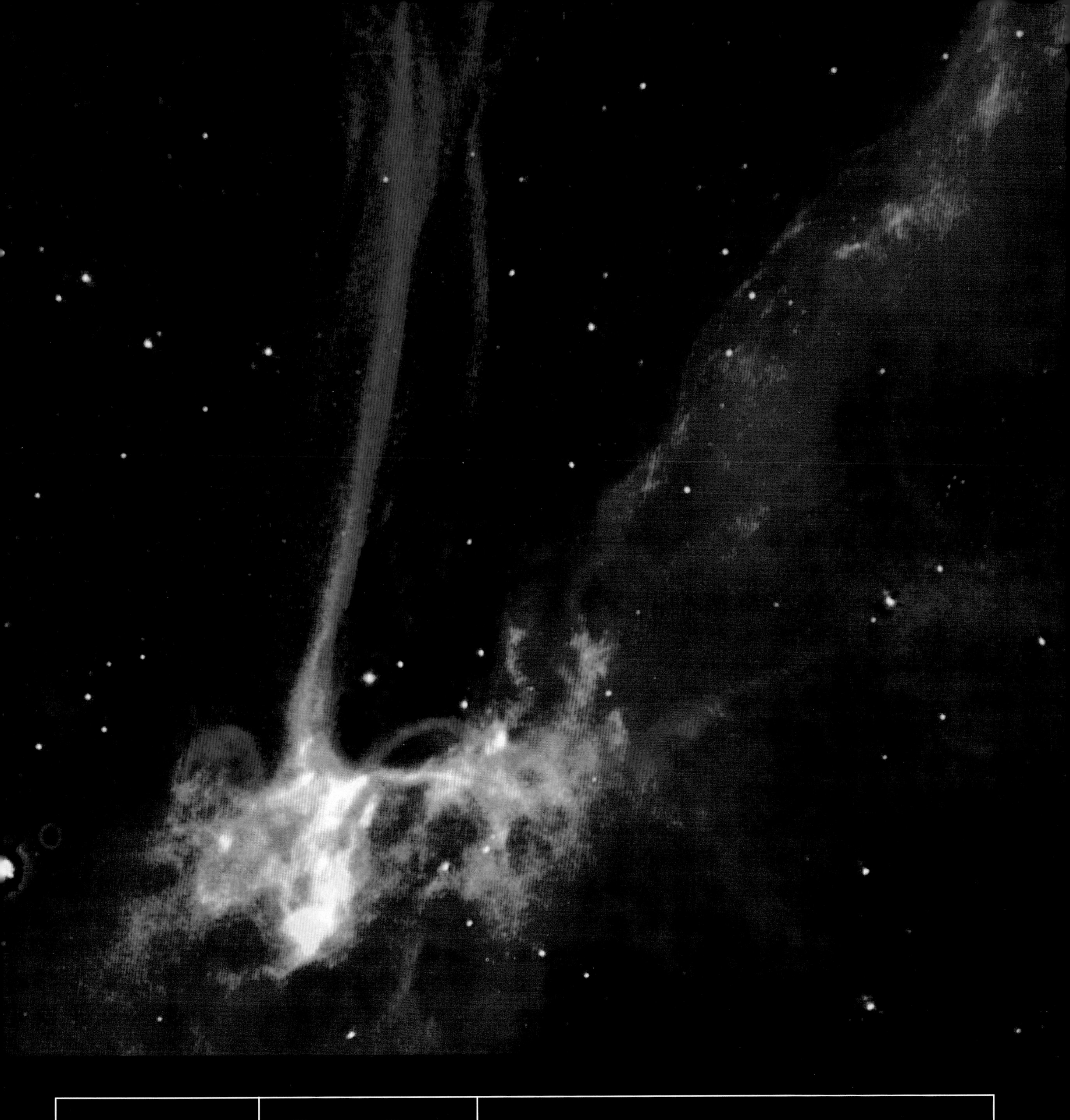

2.6 **thousand** light years	*Cygnus Loop* NGC 6960 {detail} Supernova remnant	15,000 years after it was unleashed and now spanning 80 light years, this supernova blast wave is finally slowing down – though the shock of its passage is still enough to heat the interstellar medium to over 50,000 °C (90,000 °F). The blue streak is thought to be a 5 million kph (3 million mph) bolt of gas ejected in the same explosion but only now catching the faltering wave.

3 thousand light years	*Cat's Eye Nebula* NGC 6543 Planetary nebula	The glint at the centre of this eye, as with many complex planetary nebulae, may conceal a binary system. Initially, this dying star shrugged off its outer layers in a series of regular convulsions, creating the 11 concentric rings that wrap the nebula. Then, approximately 1000 years ago, this pattern changed, inflating the multi-million degree inner bloom and outlying knots of gas.

3 **thousand** light years	*Ant Nebula* Mz 3 Planetary nebula	Another addition to our menagerie of dying stars is this light-year-sized stellar insect. No common-or-garden variety of nebula, this creature surpasses all its cousins by producing a record breaking 3.6 million kph (2.1 million mph) outflow of charged particles. Such is the spectacular diversity of planetary nebulae, one might be forgiven for eagerly anticipating our own Sun's demise.

3 thousand light years	*Sharpless 140* Star forming region	Sharpless 140's veil of dust and gas masks three newborn stars, each burning with the might of several thousand Suns. As the dust cloud around them evaporates, pungent molecules known as polycyclic aromatic hydrocarbons have been detected. So what might Sharpless 140 smell like? Probably smoke and vehicle exhaust with perhaps just a hint of burned hamburger.

3.3 **thousand** light years	*NGC 7129* Star forming region	Mayhem reigns within this molecular cloud – as it does within any nursery of energetic youngsters. This infrared head count has uncovered 130 offspring, none more than a million years old, igniting their once cold, dark parent with a blaze of radiation. Emissions reveal NGC 7129 to be rich in hydrocarbons, carbon monoxide and even water.

3.9

thousand
light years

DR 6

Star forming region

Once one star takes root within an interstellar dust cloud, others are bound to follow. At the centre of DR 6, ten heavyweight new stars have sprouted, each 10 to 20 times more massive than the Sun. Their intense heat and wind have shocked and compressed the surrounding medium, spurring further gravitational collapse within the cloud and another generation of star formation.

4 **thousand** light years	*NGC 2440* Planetary nebula	The hot white dwarf (200,000 °C or 360,000 °F) at the centre of this image has cast off its outer mantle to form this spectacular nebula. No longer fuelled by nuclear fusion, this stellar ember will slowly radiate all its heat into space eventually becoming, after many billions of years, a black dwarf. The nebula itself is unlikely to last as long – it will fade in a few millennia.

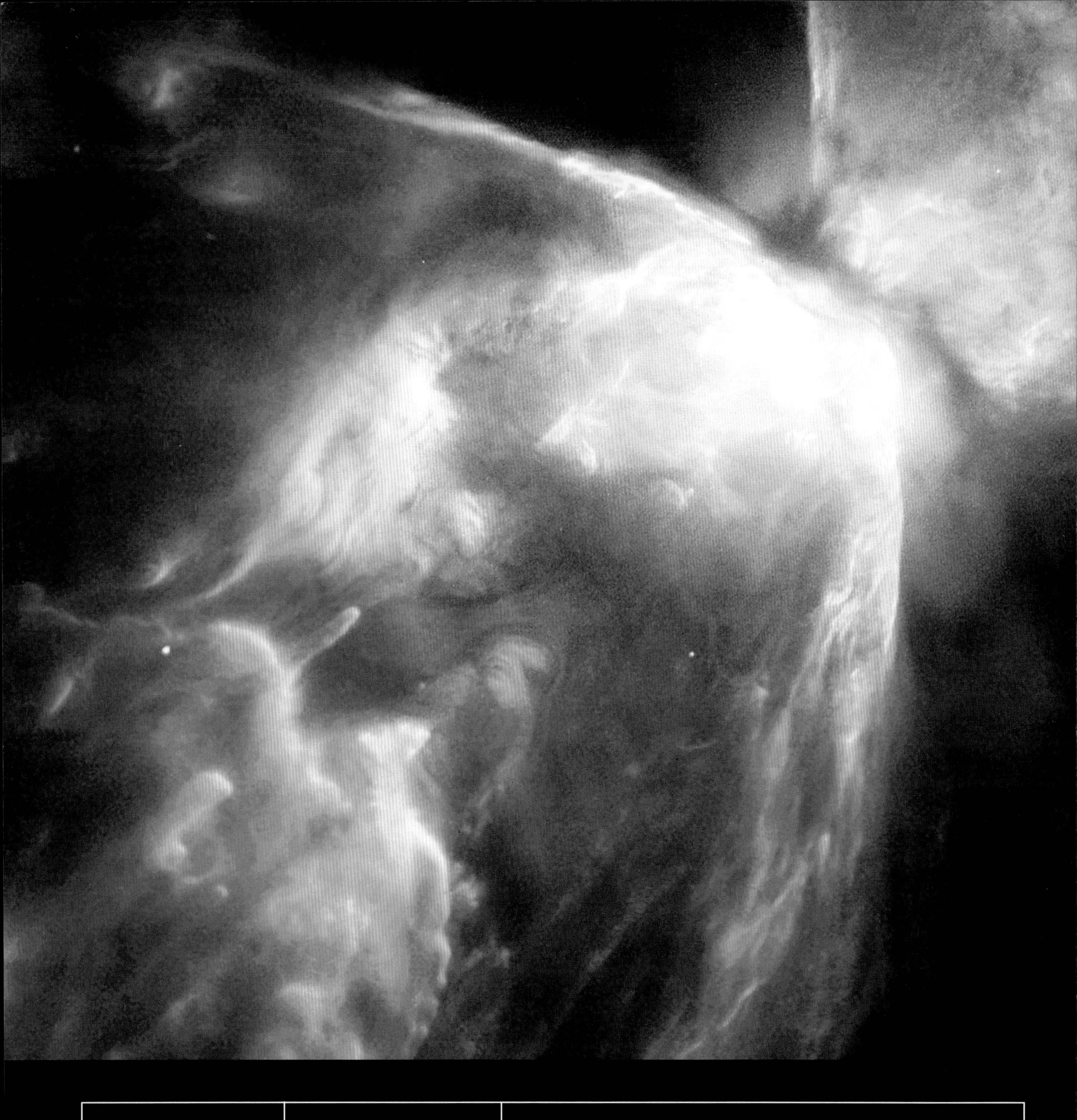

4 **thousand** light years	*Bug Nebula* NGC 6302 Planetary nebula	Residing, appropriately, in the stinger of Scorpius, the Bug Nebula is a maelstrom of fire and ice. Despite burning bright at 250,000 °C (450,000 °F), the central star is obscured by an unusual dust torus (seen top right, edge on) rich in hydrocarbons, carbonates and iron. Around these dust grains water ice has crystallized forming, in effect, a blanket of hailstones that smothers the star.

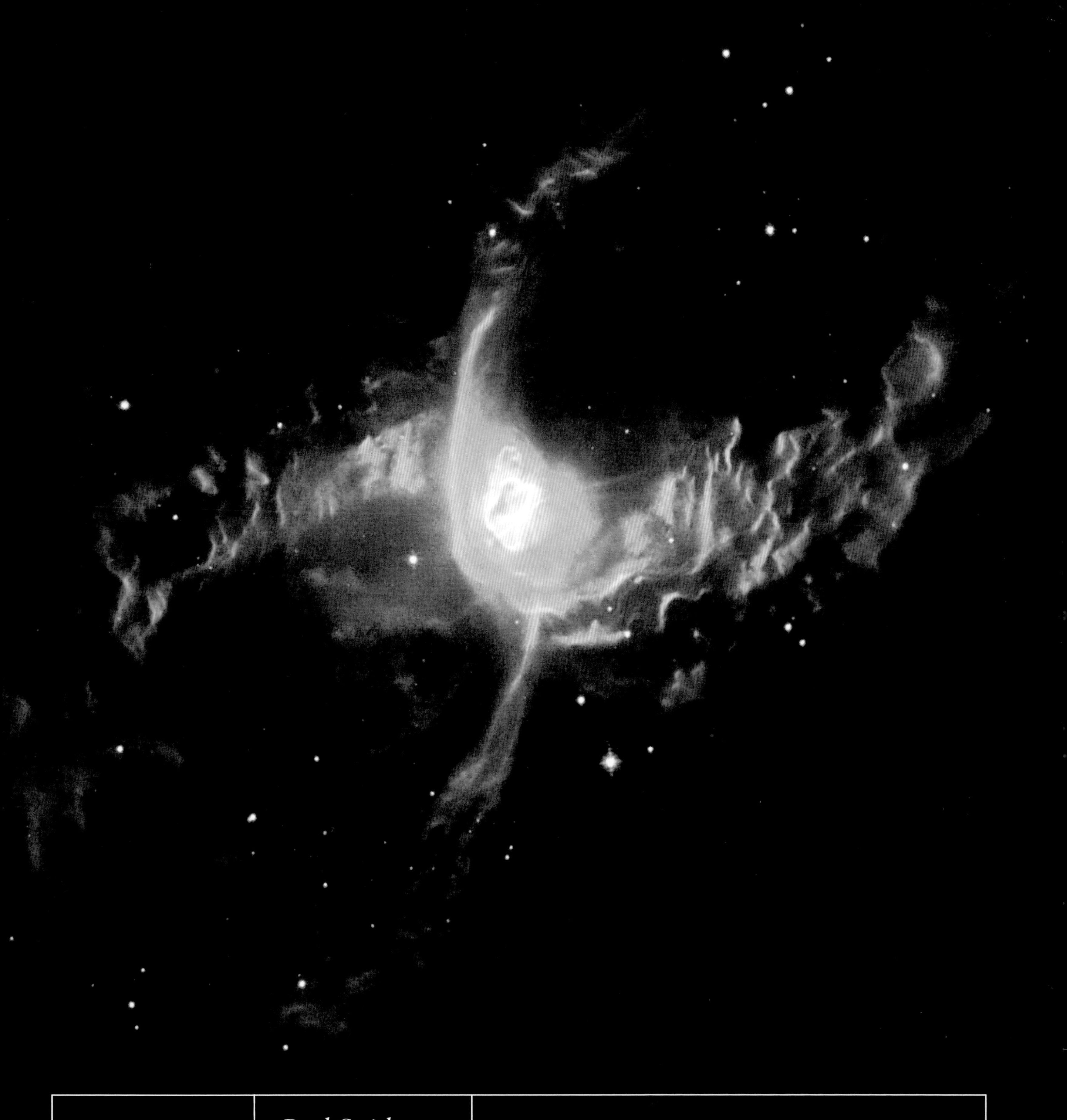

4 thousand light years	*Red Spider Nebula* NGC 6537 Planetary nebula	The stellar wind streaming from the hottest white dwarf ever observed has sculpted this cosmic arachnid. Travelling in excess of 1000 kilometres (620 miles) per second, the wind has generated waves of gas 100 billion kilometres (62 billion miles) high. The star burns at 500,000 °C (900,000 °F), while the waves are a balmy 10,000 °C (18,000 °F). Surf's up and the water's warm.

4.7

thousand
light years

Crescent Nebula NGC 6888

Emission nebula

Glowing from the shock of a furious stellar wind, this cloudscape masks the final countdown to a supernova. Blowing at 6.1 million kph (3.8 million mph), this wind emanates from a rare Wolf-Rayet class star. Super-hot and extremely short-lived (hence their rarity), Wolf-Rayets represent the final headlong rush of a massive star towards an inevitable supernova.

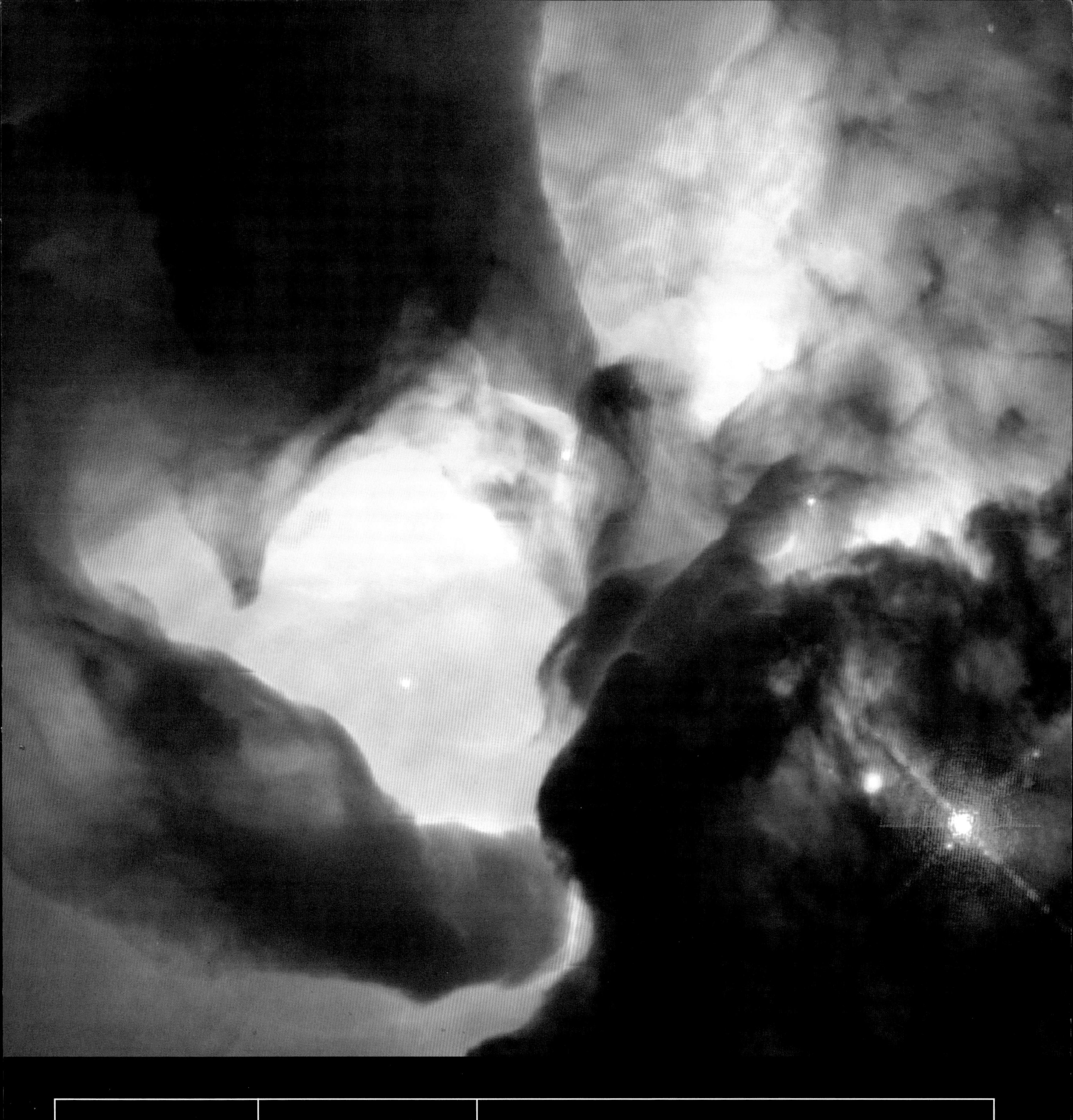

5 **thousand** light years	*Lagoon Nebula* NGC 6523 Emission nebula	A pair of one-half-light-year-long interstellar tornadoes have materialized in this prolific bed of star formation. Heated by Herschel 36 (lower right), the temperature difference between the hot surface and cold interior of the clouds, combined with strong stellar winds, may produce a strong horizontal shear that has twisted the clouds into their tornado-like appearance.

5 **thousand** light years	*Rotten Egg Nebula* OH231.8+4.2 Proto-planetary nebula	This young planetary nebula is best not approached from downwind. It owes its distinctive name to the sulphurous compounds detected in the rush of gas it is ejecting. Not only pungent, the gas is also moving at up to 1.5 million kph (900,000 mph). Its collision with the interstellar medium is charted by the two blue bubbles of shock-heated nitrogen and hydrogen seen here.

5 **thousand** light years	*Boomerang Nebula* ESO 172-07 Proto-planetary nebula	A chill wind blows through the Boomerang Nebula. At only one degree above absolute zero (-272 °C or -458 °F) it is the coldest place in the known universe – even the 13-billion-year-old afterglow of the Big Bang is warmer. Strangely, a star is responsible. This frost pocket was created by the unusually rapid 600,000 kph (400,000 mph) expansion of material ejected by a red giant.

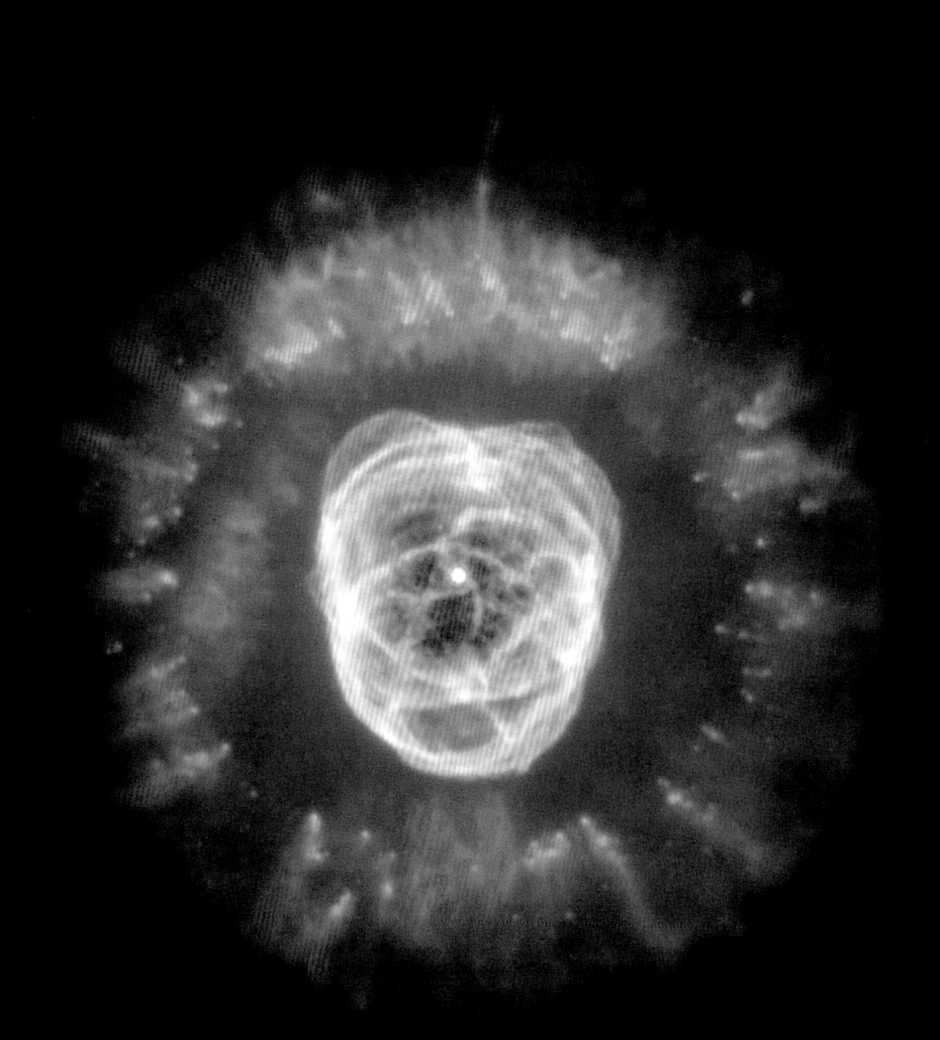

5 **thousand** light years	*Eskimo Nebula* NGC 2392 Planetary nebula	This Eskimo's parka disguises another bipolar planetary nebula. Its second lobe is concealed directly behind this one – we are viewing the nebula edge on. The parka's orange fur trim is thought to be formed by slow-moving globules of gas streaming in an eroding flow of faster-moving material. And it is moving quickly: this Eskimo's hood is growing at 1.5 million kph (900,000 mph).

5.5

thousand
light years

RCW 38

Star forming region

To our eyes RCW 38 is another dark molecular cloud but seen in infrared it opens to reveal a treasure chest of new stars. Embedded within the cloud are a number of hypergiant O-class stars. At least 20 times heavier than the Sun and 35,000 times as luminous, O-stars live brief lives of no more than six million years indicating that star formation is an on-going process in RCW 38.

thousand
light years

Swan Nebula

M17 {detail}

Star forming region

Fluorescing in the heat of newly smelted stars, excited atoms of hydrogen, nitrogen, oxygen and sulphur have woven a rainbow of light across the Swan Nebula's dense mists. The presence of elements heavier than hydrogen reveals the Swan to be both nursery and graveyard; a crucible in which the ashes of earlier stellar generations are re-forged to create the next.

5.5

thousand
light years

Swan Nebula

M17 {detail}

Star forming region

Storm clouds glowing with hydrogen and sulphur gather against a blue oxygen sky. Roughly three light years across, these turbulent forms have been sculpted and set ablaze by a torrent of ultraviolet radiation from young, massive stars, lying outside the picture to the upper left. Such vigorous remodelling may spark a new wave of stellar formation within the cloud.

5.9 **thousand** light years	*IC 2944* Bok globules	Particularly dense knots of interstellar dust and gas, Bok globules are the swaddling clothes of star formation. The largest globule (top right) is actually two overlapping clouds. Each cloud spans 1.4 light years and contains enough material to make seven Suns. But these globules will never bear stars, they are evaporating in the glare of the cluster of supergiants seen here.

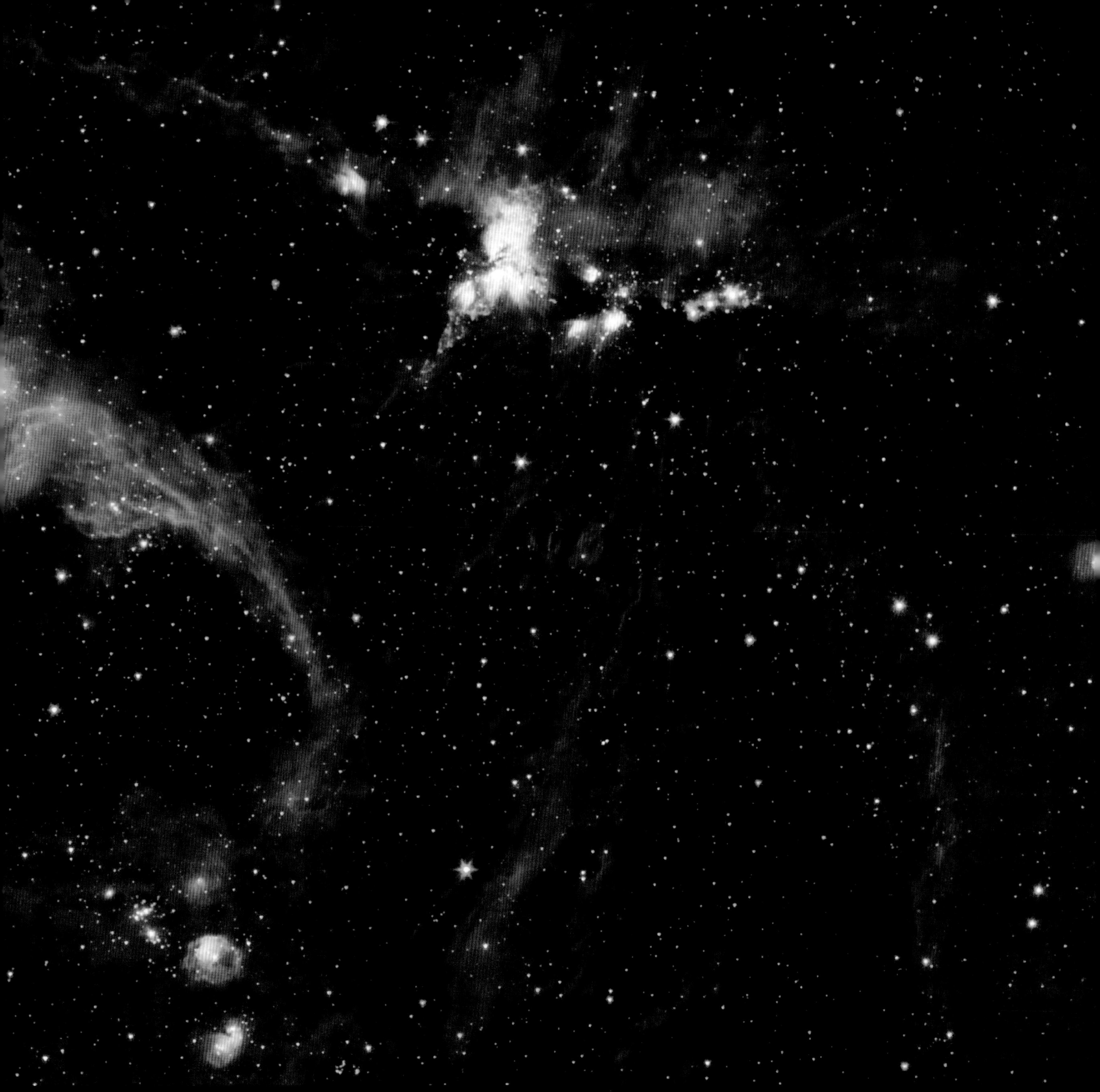

6.2 **thousand** light years	*DR 21* Star forming region	Behind a veil of dust so thick only infrared light can escape, DR 21 cradles a newborn star 100,000 times brighter than our Sun. This heavyweight sits at the centre of a web of tangled filaments shocked out of the surrounding molecular cloud by a massively powerful jet of plasma (seen here in green) and gale force stellar winds.

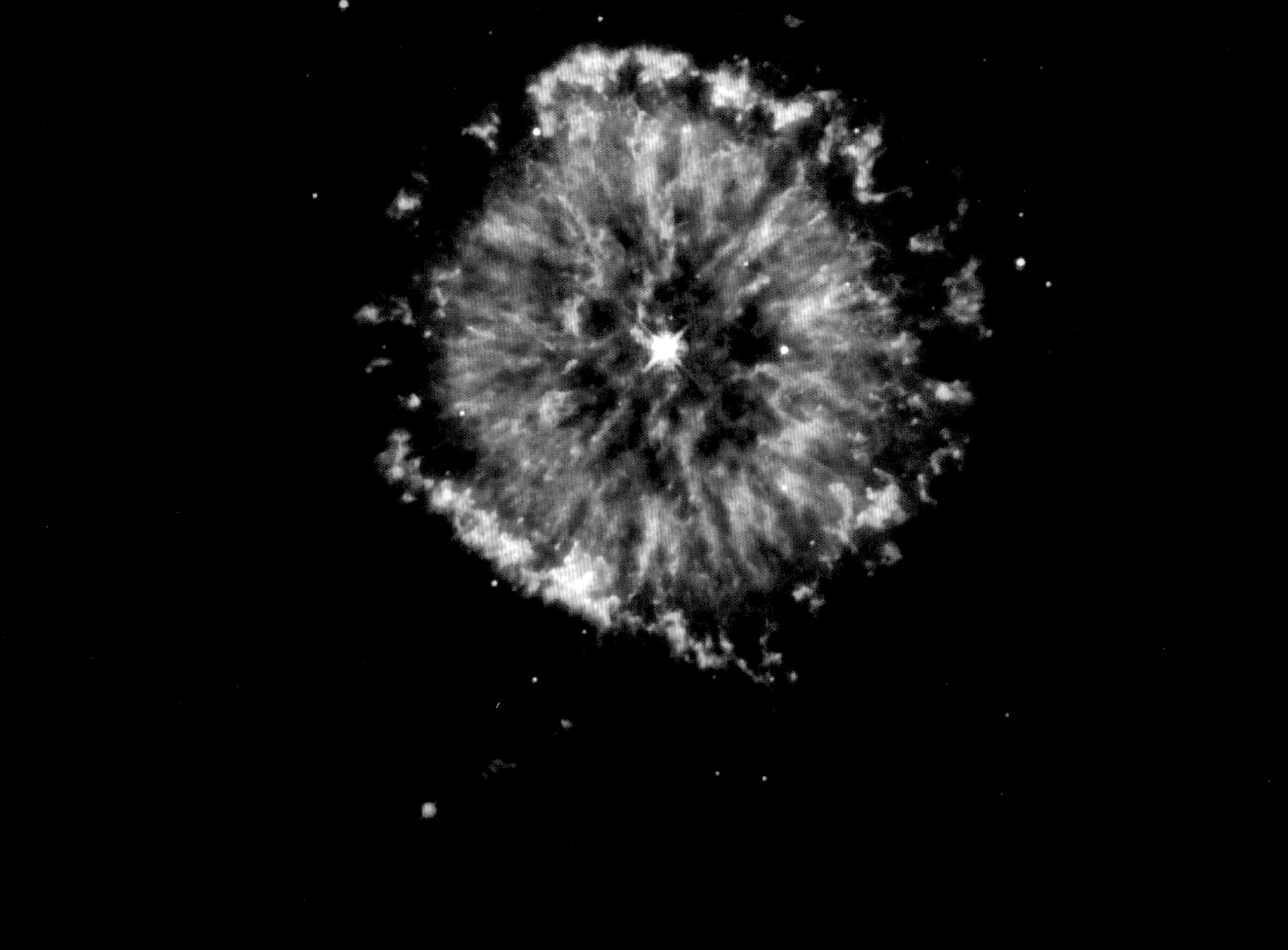

6.5 **thousand** light years	*NGC 6751* Planetary nebula	Lighting this flaming eye is a ball of exotic matter. Up to a million times denser than the star it formed from, the crush of gravity on a white dwarf is so extreme it compresses atoms themsleves, forcing electrons towards the nucleus. Saved from further collapse only by quantum laws that forbid electrons from sharing the same energy state, matter in this form is said to be 'degenerate'.

6.5 **thousand** light years	*Gomez's Hamburger* Proto-planetary nebula	Here lies a star poised between giantism and dwarfdom. Outer layers shed as the last drops of nuclear fuel were burned have fashioned a stifling belt of dust so thick light can only escape above and below it, creating the 'bun' we see here. Now spent, the remains of the star will collapse into a white dwarf, heating the surrounding belt till it glows as a fully fledged planetary nebula.

6.5

thousand
light years

Crab Pulsar

PSR 0531+21

Pulsar

Forged by the supernova that lit up Earth's skies in 1054 AD, the Crab Pulsar shines as brightly as the Sun, despite being less than 30 km (20 miles) across. At x-ray wavelengths, it produces 100 times as much energy as the Sun, accelerating particles close to the speed of light and flinging them out into space, energizing and illuminating the nebula that extends for 10 light years around it.

7

thousand
light years

Eagle Nebula

M16

Dark & emission nebula

Hewn from a dense cloud of hydrogen by a torrent of plasma, these iconic stalagmites are replete with compact pockets of gas dubbed Evaporating Gas Globules or EGGs. Within many EGGs, protostars are quickening in a race against time as their sustaining gas yolk is boiled away by photoevaporation. Astronomers are still unsure which came first: the star or the EGG?

7.1

thousand
light years

Bubble Nebula

NGC 7635

Emission nebula

This stormy bubble of gas is being inflated by a stellar wind moving at 7 million kph (4.5 million mph), blowing from a hypergiant star 40 times more massive than the Sun and over 400,000 times as luminous. Even beyond the storm's edge, the hypergiant's influence is still felt – the gas tendrils at the top of the picture are eroding in its glare.

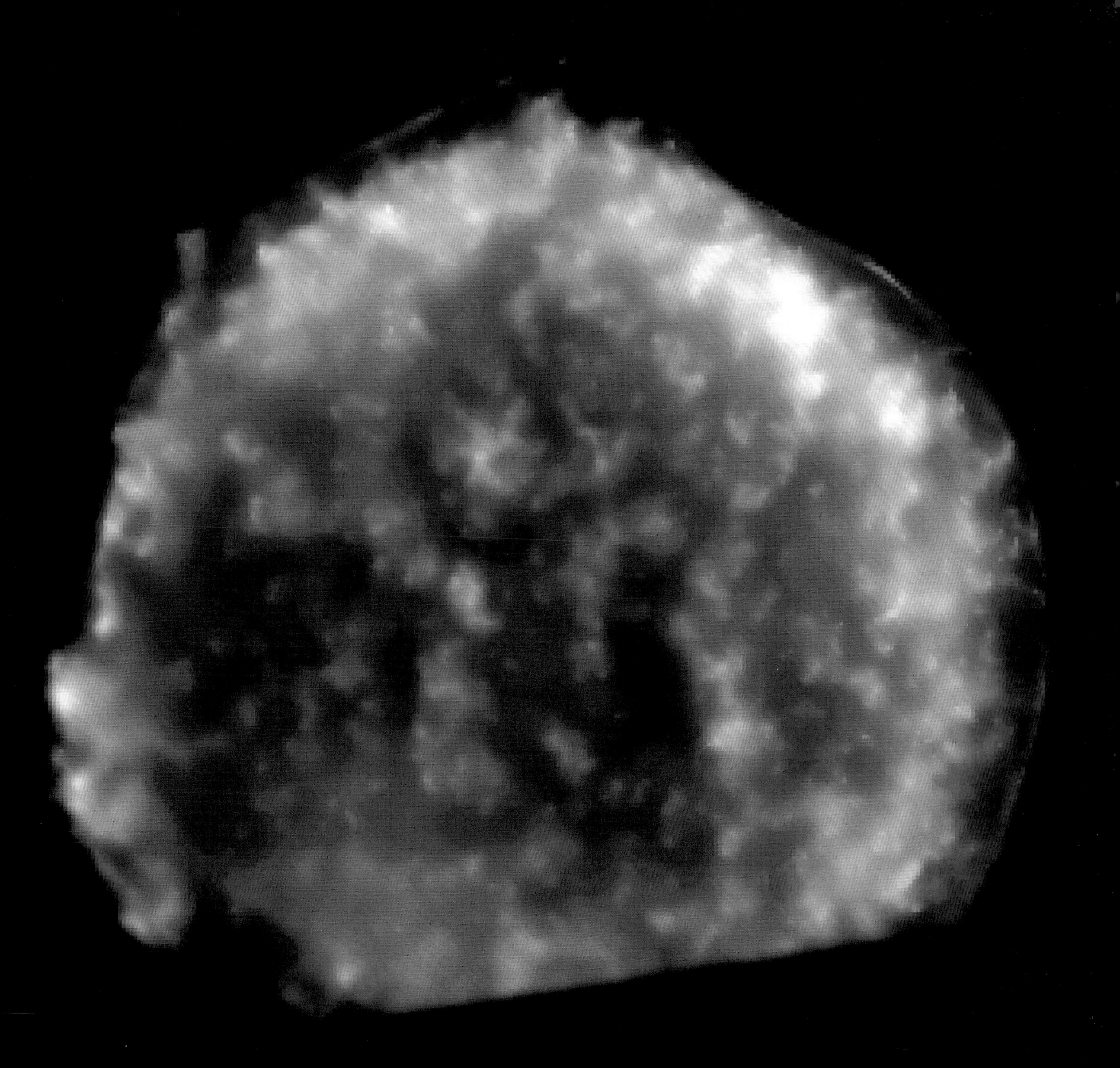

7.5

thousand
light years

Tycho's SNR

SNR 1572

Supernova remnant

White dwarfs can come back from the dead by devouring their neighbours. If enough mass is consumed a wave of fusion re-ignites the star as the most powerful of all supernovae, a Type 1a. These detonations utterly destroy their hosts, leaving no central neutron star or black hole. Such was the event witnessed by Tycho Brahe in 1572; its debris now creates this stellar pollen grain.

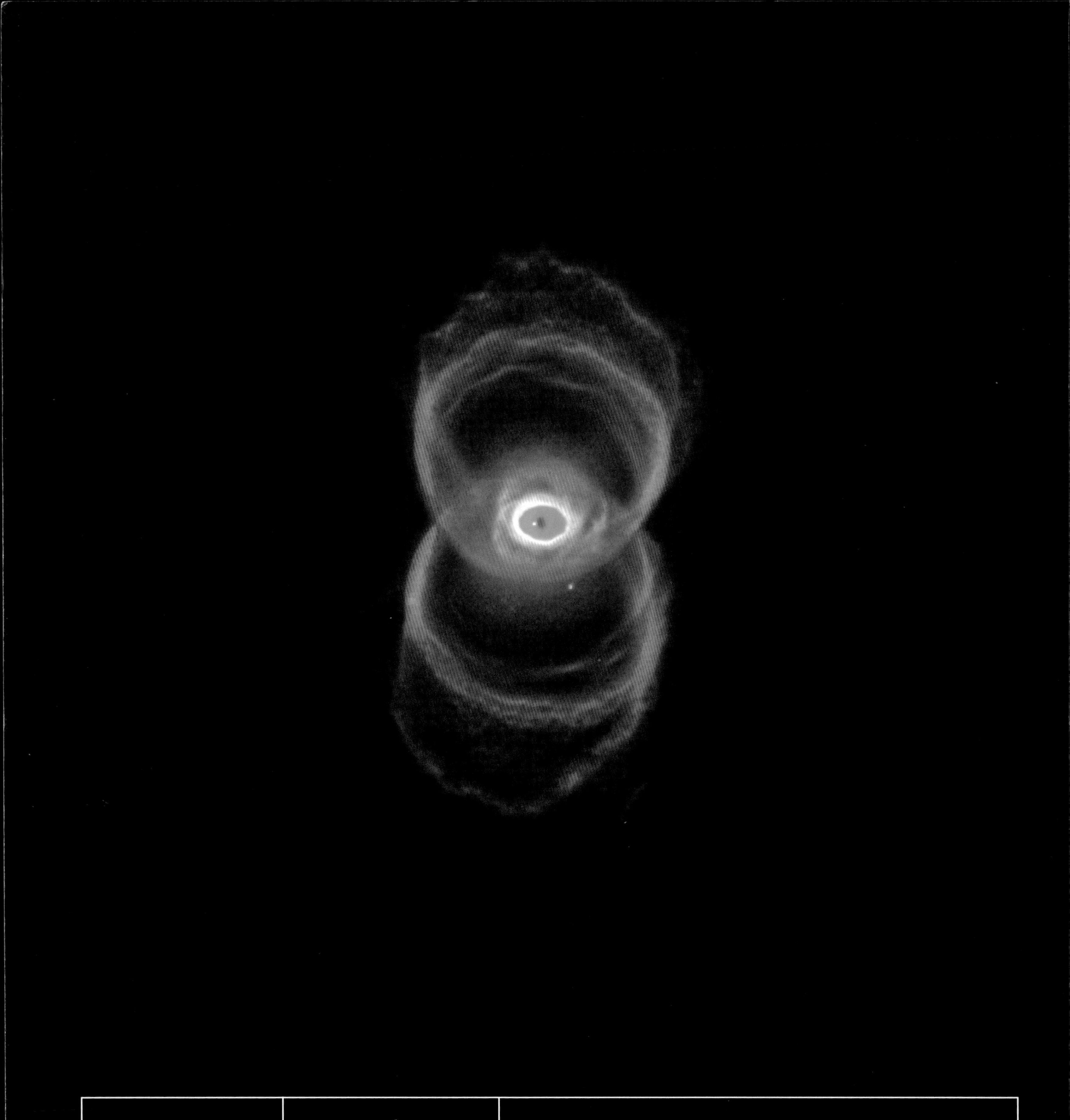

8

thousand
light years

Hourglass Nebula MyCn18

Planetary nebula

Burning bright, the fearful symmetry of the Hourglass Nebula is framed by a process as yet unknown. Explanations of its complexity range from mass ejection along magnetic field lines to gravitational interaction with an unseen companion. As the Hourglass appears to contain too much material to be forged from just one star, the second hypothesis seems more likely.

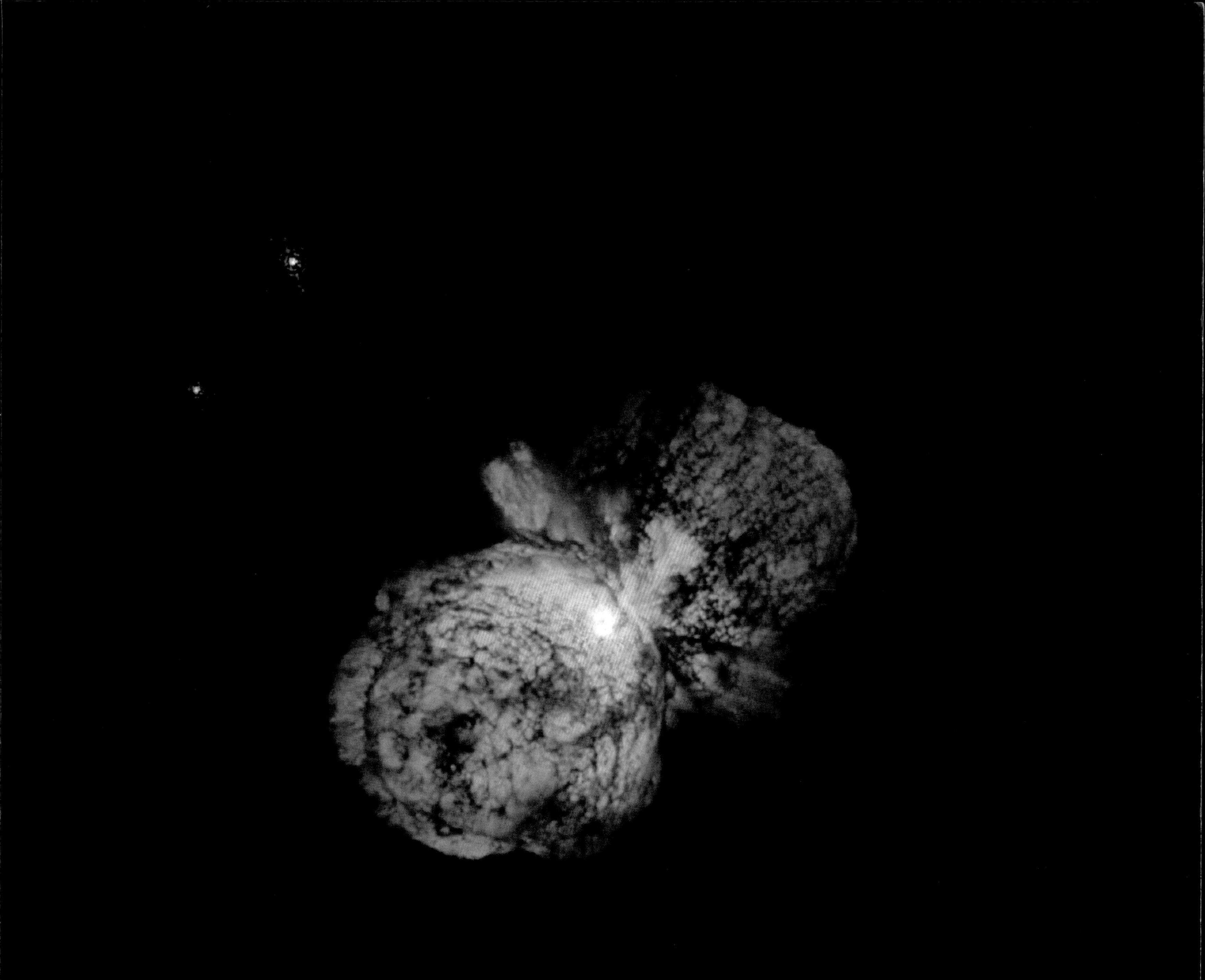

8 thousand light years	*Eta Carinae* Star	Hypergiants live fast and die young and few stars come more hyper or giant than Eta Carinae. 100 times more massive than the Sun and 4 million times brighter, Eta Carinae is living on borrowed time. It survives only by ejecting vast quantities of matter with near-supernova force, creating the 12 trillion kilometre (8 trillion mile), 60 million °C (108 million°F) nebula that surrounds it.

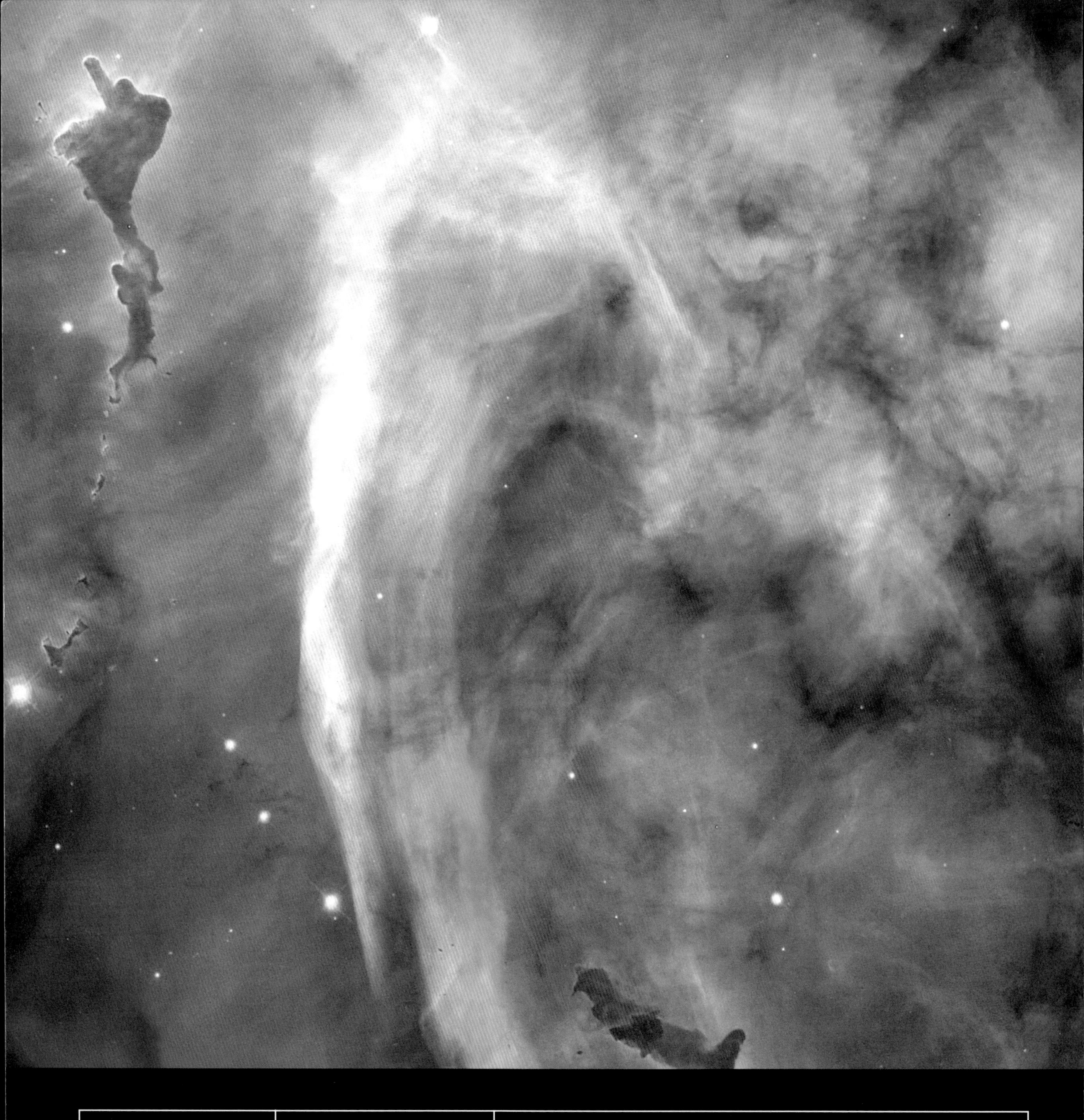

8 **thousand** light years	*Keyhole Nebula* NGC 3372 Star forming region	Lying at the heart of the Carina Nebula, the Keyhole structure churns with bright filaments of hot, fluorescing gas and sharply silhouetted clouds of cold molecules and dust. Like the Eagle Nebula's famous pillars of creation, the leftmost of these clouds is evaporating in the glare of a nearby star, exposing burgeoning globules of star forming material.

8.2 **thousand** light years	*NGC 6397* Globular cluster	In this stellar hive, stars swarm in densities a million times greater than in our neighbourhood. Often separated by mere light weeks, they run a very real risk of collision. Where collisions do occur, the two (or more) stars merge to form a 'blue straggler', a new, hotter, brighter star that stands out against the cluster's aging population. Several such stragglers appear in this image.

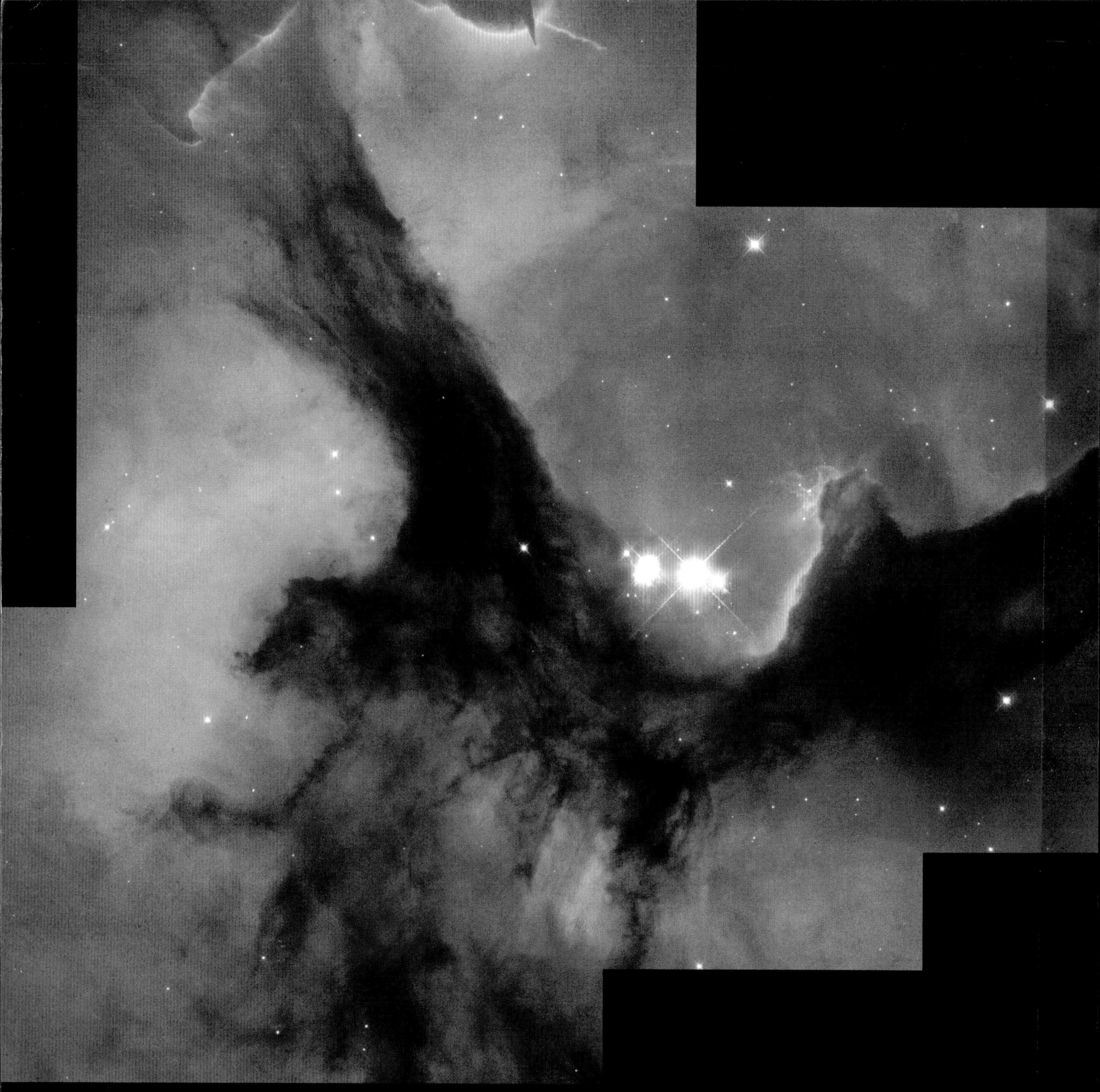

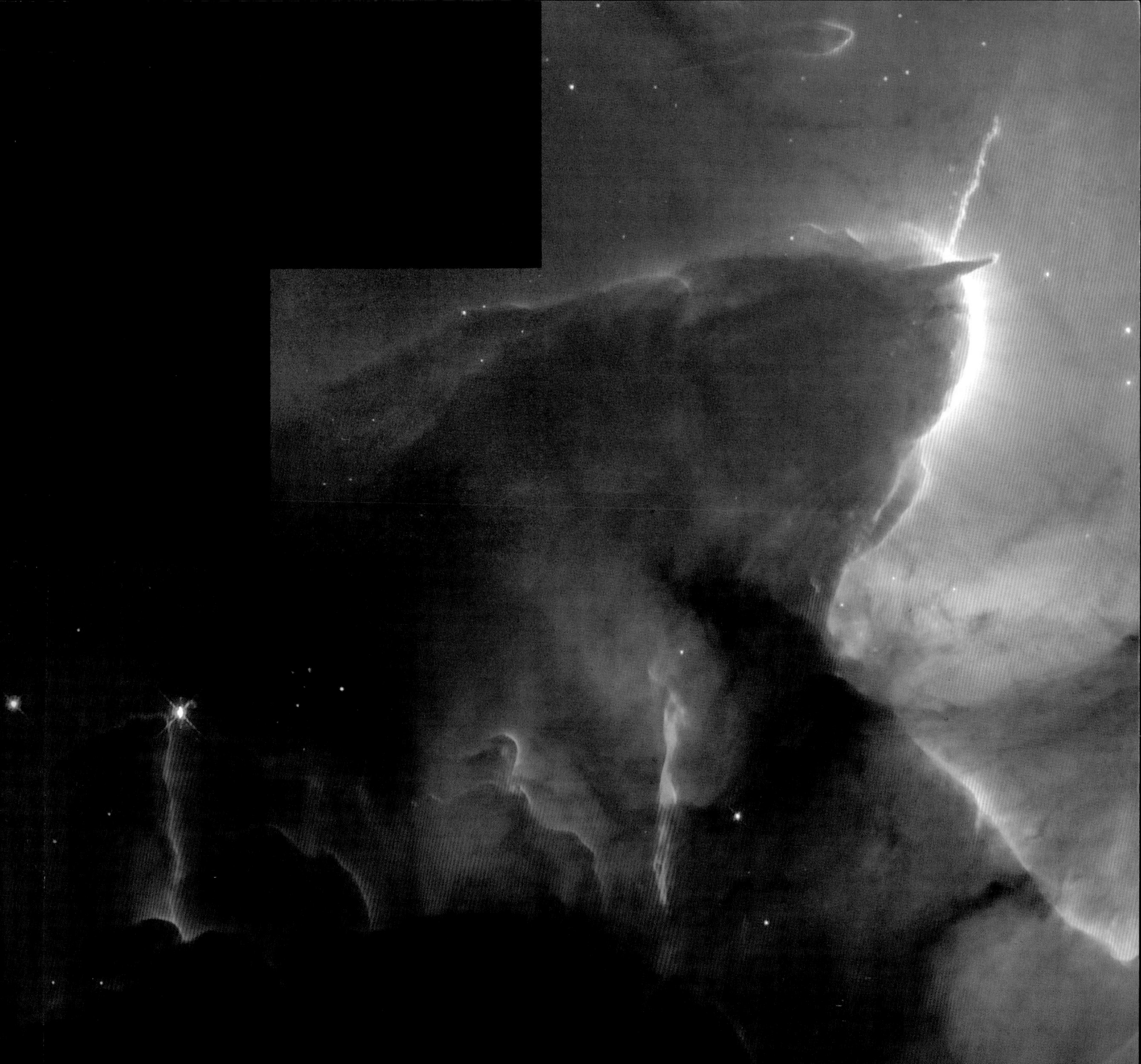

9 **thousand** light years	*Trifid Nebula* NGC 6514 {detail} Star forming region	O-class supergiants brook no competition. Scourging the nebula with radiation, they conspire to stunt their younger siblings by evaporating the clouds in which they grow. Deprived of sustenance, these infants will now form lower-mass stars like our Sun rather than supergiants. They may never outshine their elder brothers, but they will outlive them many times.

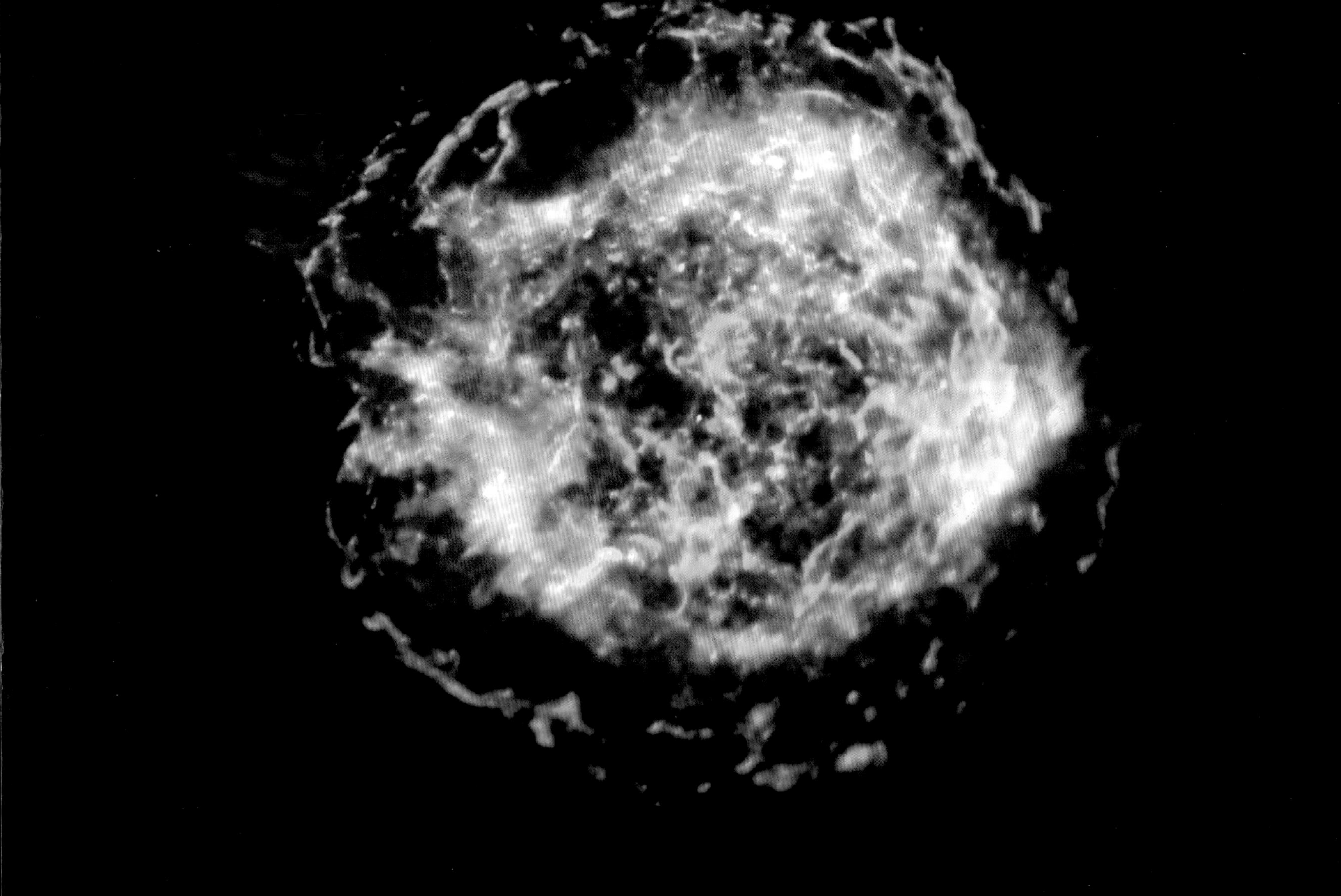

10 **thousand** light years	*Cassiopeia A* Supernova remnant	Without the alchemy of stars we could not exsist. Their lives enrich the sparse hydrogen and helium soil of the cosmos creating carbon, silicon and iron while the final multi-billion degree furnace of a supernova forges the universe's heavier elements. Here, X-ray vision reveals two distinct jets of silicon (red) and iron (blue) in the elemental harvest of supernova remnant Cassiopeia A.

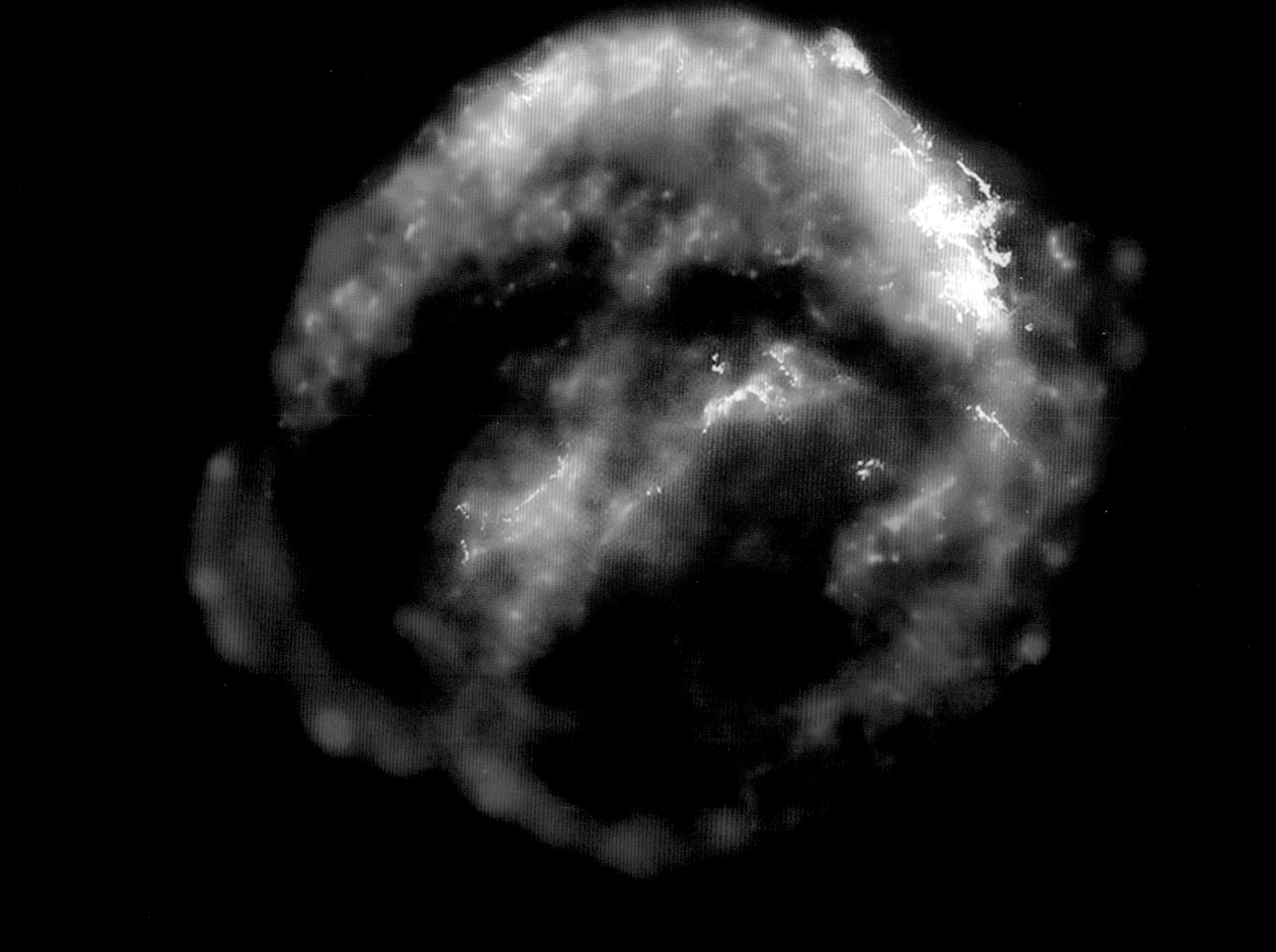

13 **thousand** light years	*Kepler's SNR* SN 1604 Supernova remnant	Now 14 light years wide, the shockwave from the supernova that lit the skies above Johannes Kepler in 1604 is still tearing through space at 6 million kph (4 million mph). No-one is sure what sparked the explosion – was it the inevitable fate of a massive star (a Type II supernova) or the all too brief resurrection of a white dwarf (a Type I supernova)?

13.7 **thousand** light years	*RCW 49* Star forming region	350 light years across and housing over 2,200 stars, RCW 49 is one of our galaxy's most industrious stellar construction grounds. At its centre production has now halted – a cluster of hefty supergiants have burned away the dusty haze they condensed from – but deeper in the nurturing clouds some 300 protostars have been identified, several with planetary discs.

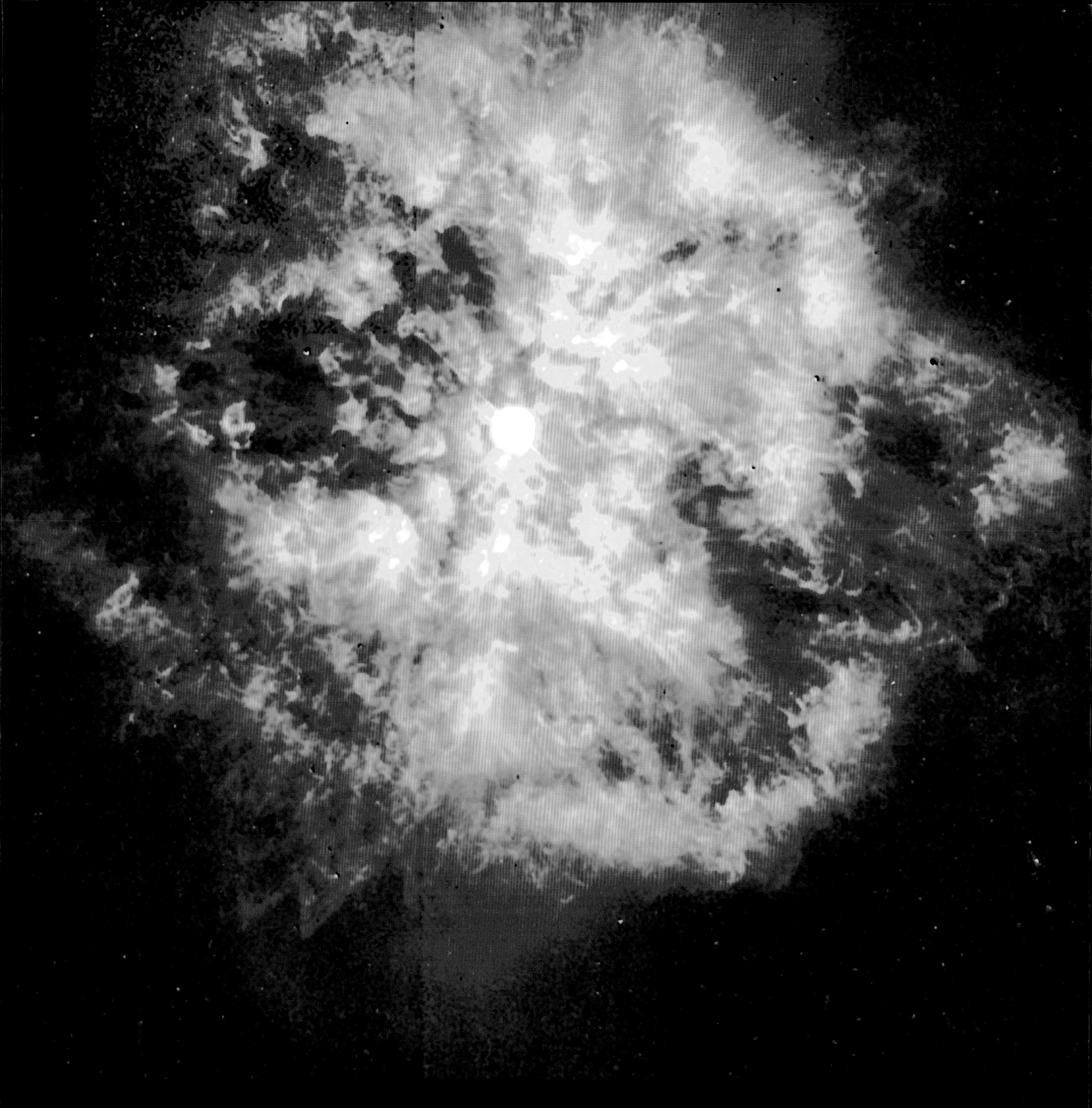

15 **thousand** light years	*WR 124* Wolf-Rayet star	Wolf-Rayet stars are poised on the edge of supernova – WR 124 may look as if it has already exploded, but the main event is still to come. An extreme stellar wind 10 billion times fiercer than our Sun's is flaying this star alive, tearing 160 billion kilometre (100 billion mile) clumps of matter from its surface. Such prodigious mass loss will only hasten its inevitable fate.

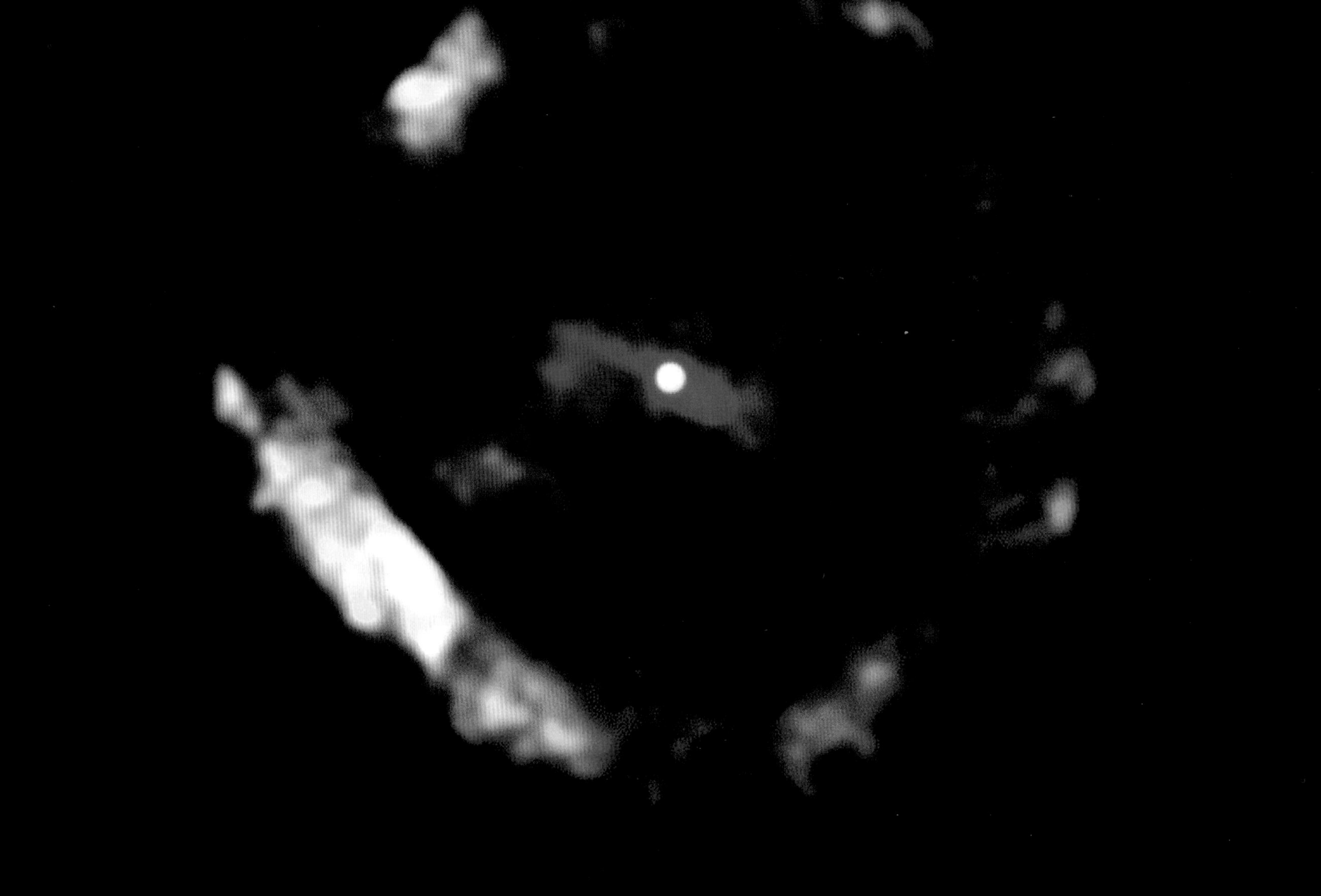

16 **thousand** light years	*G11.2-0.3* Pulsar	Pulsars don't actually pulse – they spin. Like a ship at sea swept by a light-house's beam, we perceive the pulsar to flash on and off as its constant rays sweep past us. A supernova witnessed by Chinese astrologers in 386 AD set this neutron star spinning at a relatively leisurely 14 times a second – the fastest pulsar on record spins at over 600 times a second.

17
thousand
light years

Omega Centauri NGC 5139

Globular cluster

With a population numbering millions, this glittering metropolis is Omega Centauri, our galaxy's largest globular cluster. In this image 50,000 stars are packed into a 13-light-year area – a similar plot centred on our galactic suburb would hold just six. It is possible that clusters like these are the remnant nuclei of small galaxies swallowed whole by the Milky Way.

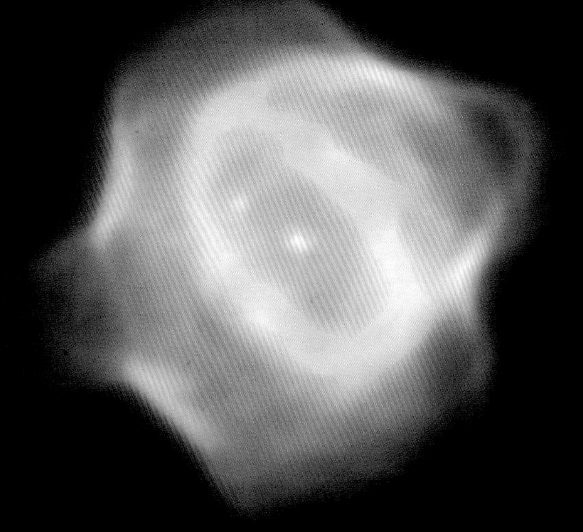

18 **thousand** light years	*Stingray Nebula* Hen-1357 Planetary nebula	The Stingray has only just emerged from the dark. Twenty years ago, the nebulous gas entombing the dying star wasn't hot enough to glow. As the Stingray is switching on, its central star is switching off. The intense radiation needed to illuminate the nebula is generated not by fusion, but by the star collapsing under its own weight as its nuclear furnaces finally go out.

20

thousand
light years

NGC 3603

Emission nebula

Amongst the clouds of the galaxy's largest nebula, the stellar lifecyle is revealed in its entirety: glowing pillars of hydrogen herald the birth of newborns; mature blue giants cluster in a clearing scorched from the cloud they formed in; aging star Sher 25 blows smoke rings as it lurches towards a supernova that will seed the skies with heavy elements and trigger a new wave of creation.

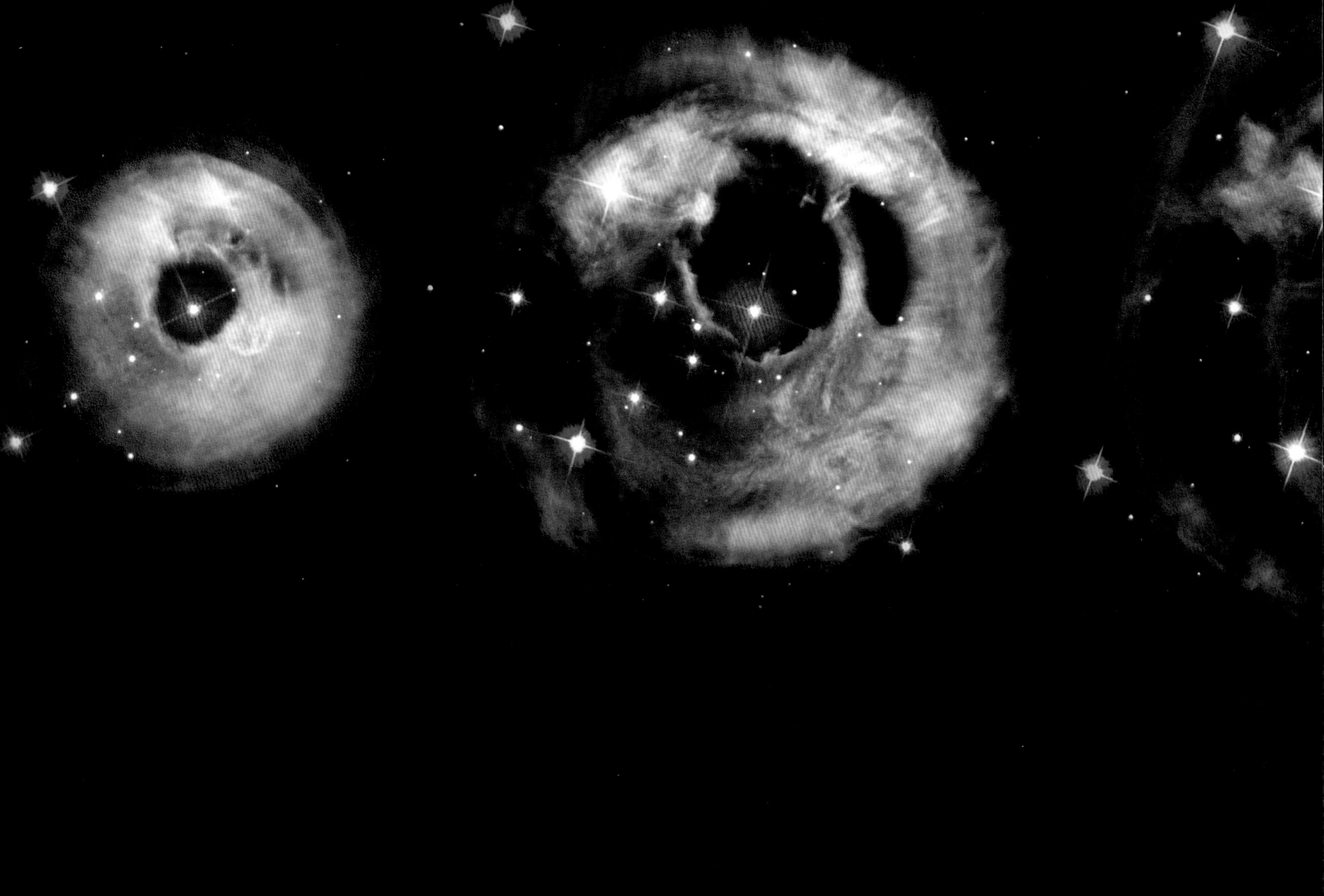

20 **thousand** light years	*V838 Monocerotis* Star	In January 2002, a faint star on the outer edges of the Milky Way flared to 600,000 times the luminosity of the Sun. By April the flare had subsided but its echoes were still reaching Earth years later. This remarkable timelapse bloom is conjured from light bounced back from ever wider shells of debris ejected by V838 Monocerotis during a previous outburst.

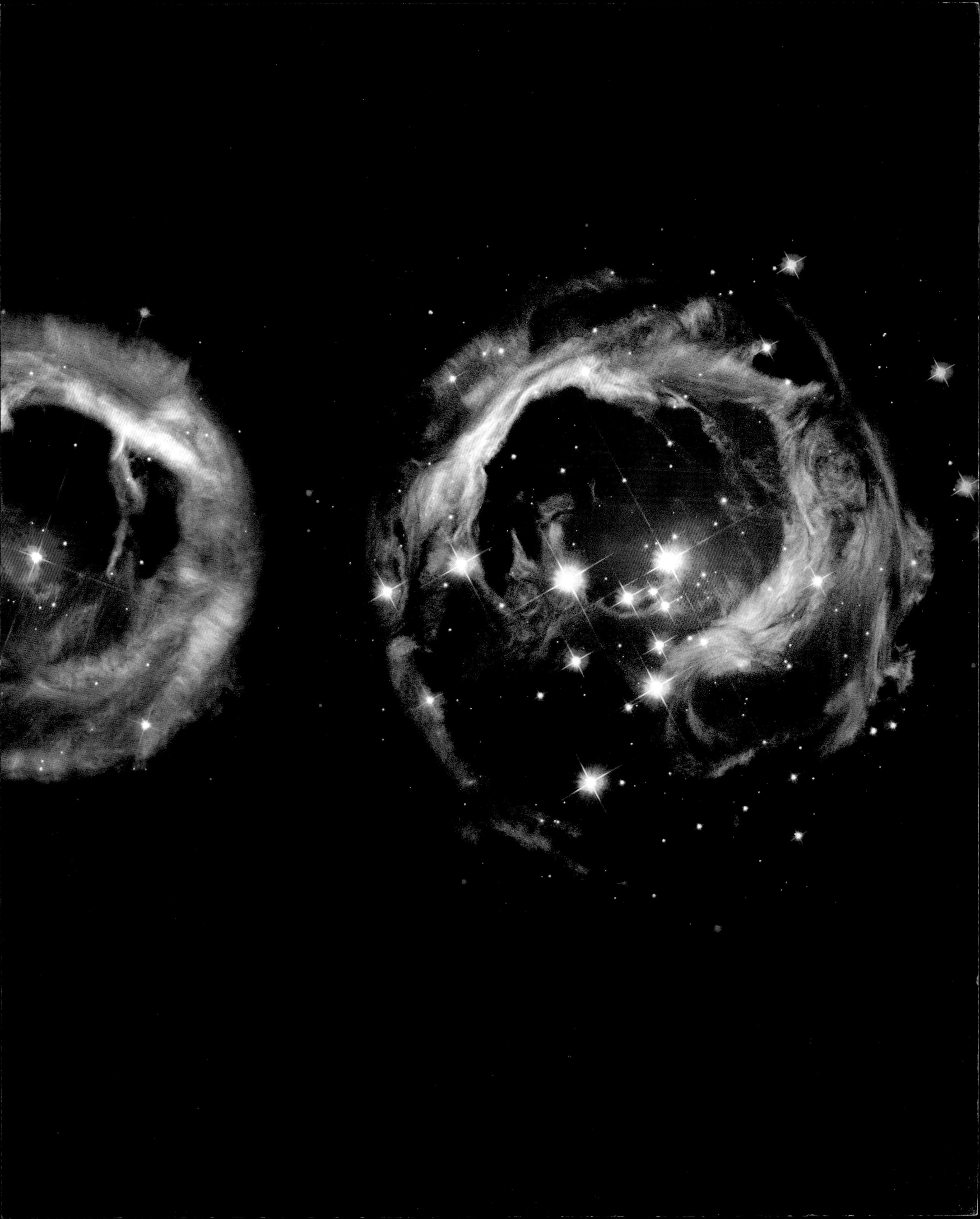

25
thousand
light years

Quintuplet Cluster

Open star cluster

We now close to within 100 light years of the galactic core, an extreme environment that begets extreme stars. This young cluster has spawned the galaxy's densest gathering of stellar titans, presided over by the greatest of them all, the Pistol Star (below centre). Over the next million years this parliament of giants will dissolve into the gravitational maelstrom that surrounds them.

25

thousand
light years

Pistol Star

G0.15-0.05

Star

The Pistol Star is large enough to fill Earth's orbit and is thought to release as much energy in 1.25 seconds as the Sun does in a year, making it 25 million times as luminous. It is the biggest, brightest star known – and it used to be bigger: it's so hot that gravity can't keep it together and it has ejected perhaps 50 percent of its original mass, creating the nebula we see around it.

25
thousand
light years

Milky Way Centre

Galactic core

Dense star fields (should you care to count them, there are nearly 10 million stars here) announce our arrival at the galactic core. In these bright depths galactic orbits are completed in decades – it takes our Sun 225 million years. A ribbon of dust weaves diagonally across the page leading to the exact point around which we all revolve (top right), where beats our galaxy's dark heart.

25

thousand
light years

*Sagittarius A**

Supermassive black hole

A blaze of x-ray radiation (right of centre) uncloaks Sagittarius A*, the supermassive black hole that dominates the heart of the Milky Way. Weighing in at 3 million solar masses, Sagittarius A* holds court over a melting pot of white dwarves, supernovae, neutron stars and black holes, all bathed in an incandescent fog of multimillion-degree gas.

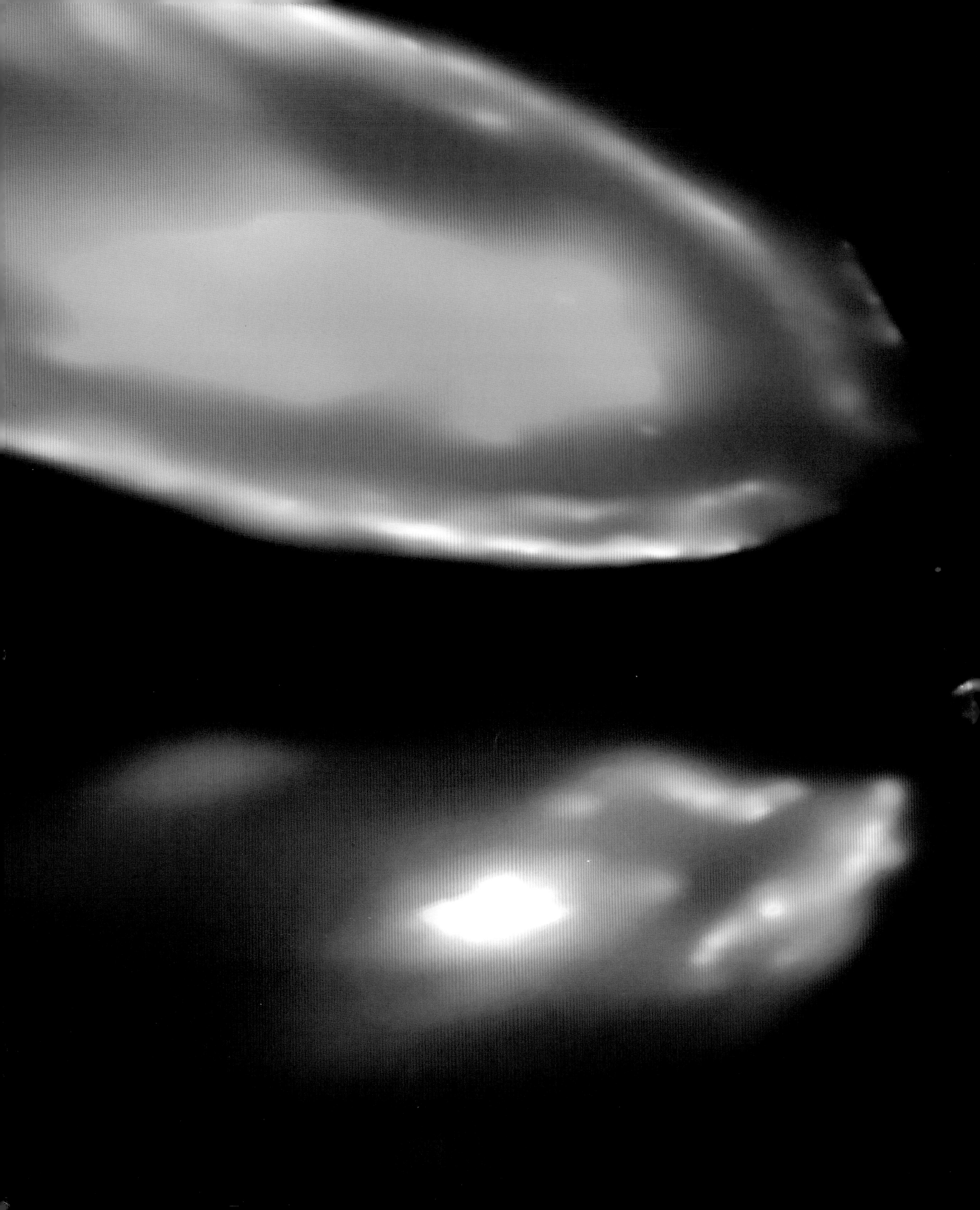

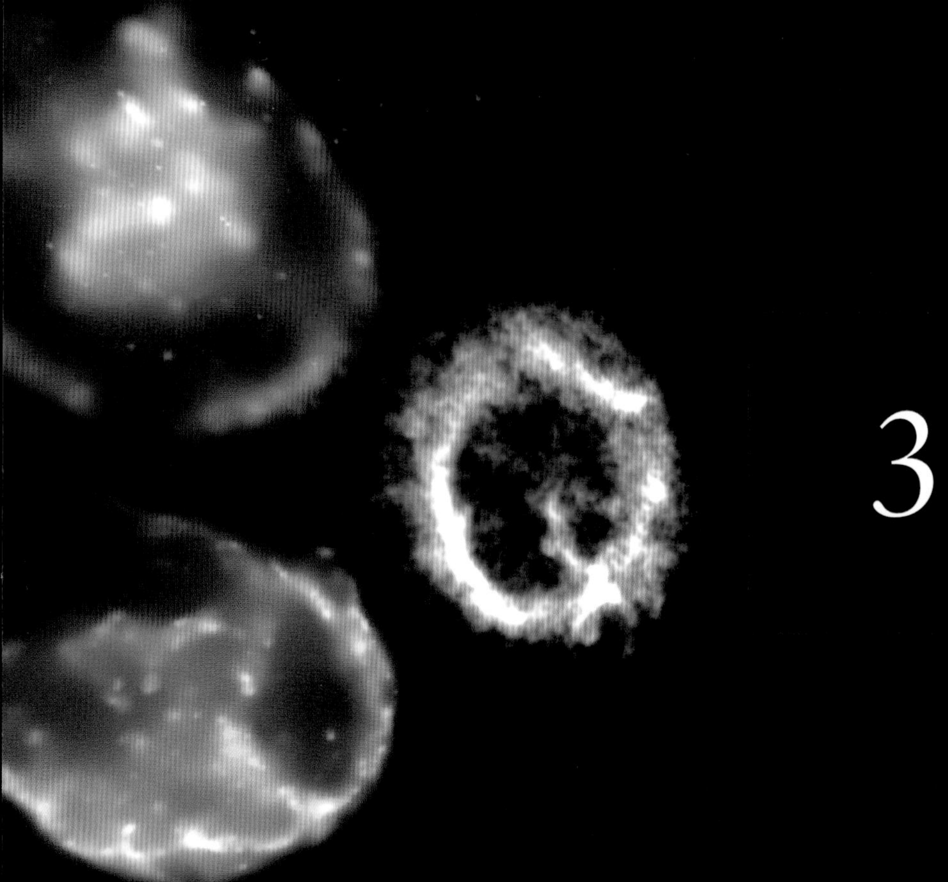

3 magellanic clouds

160

thousand
light years

Large Magellanic Cloud Irregular galaxy

Named after Ferdinand Magellan who identified them as he circumnavigated the globe, the Large and Small Magellanic Clouds are satellite galaxies gravitationally bound to the Milky Way. Ultimately destined to be consumed by their larger neighbour, for now they are close members of the 'Local Group', a collection of 40 galaxies occupying a sphere 10 million light years across.

160 **thousand** light years	*N11B* Star forming region	Hot stars, abrasive winds and violent radiation sculpt pillars of dust and mountains of gas in one of the most fertile star forming regions of the Large Magellanic Cloud. Second only to the Trifid Nebula in terms of star production, N11B spans 100 light years and houses some of the most massive stars in the known universe (top left quadrant).

160 thousand light years	*N49* Magnetar	A novel species of neutron star is concealed within these filaments of supernova debris. Dubbed a magnetar, it boasts a magnetic field a thousand trillion times stronger than Earth's. On such a star a lump of iron would be subject to a force 150 million times stronger than Earth's gravity. There is a price to pay for such exuberance: a magnetar's lifespan is measured in thousands of years.

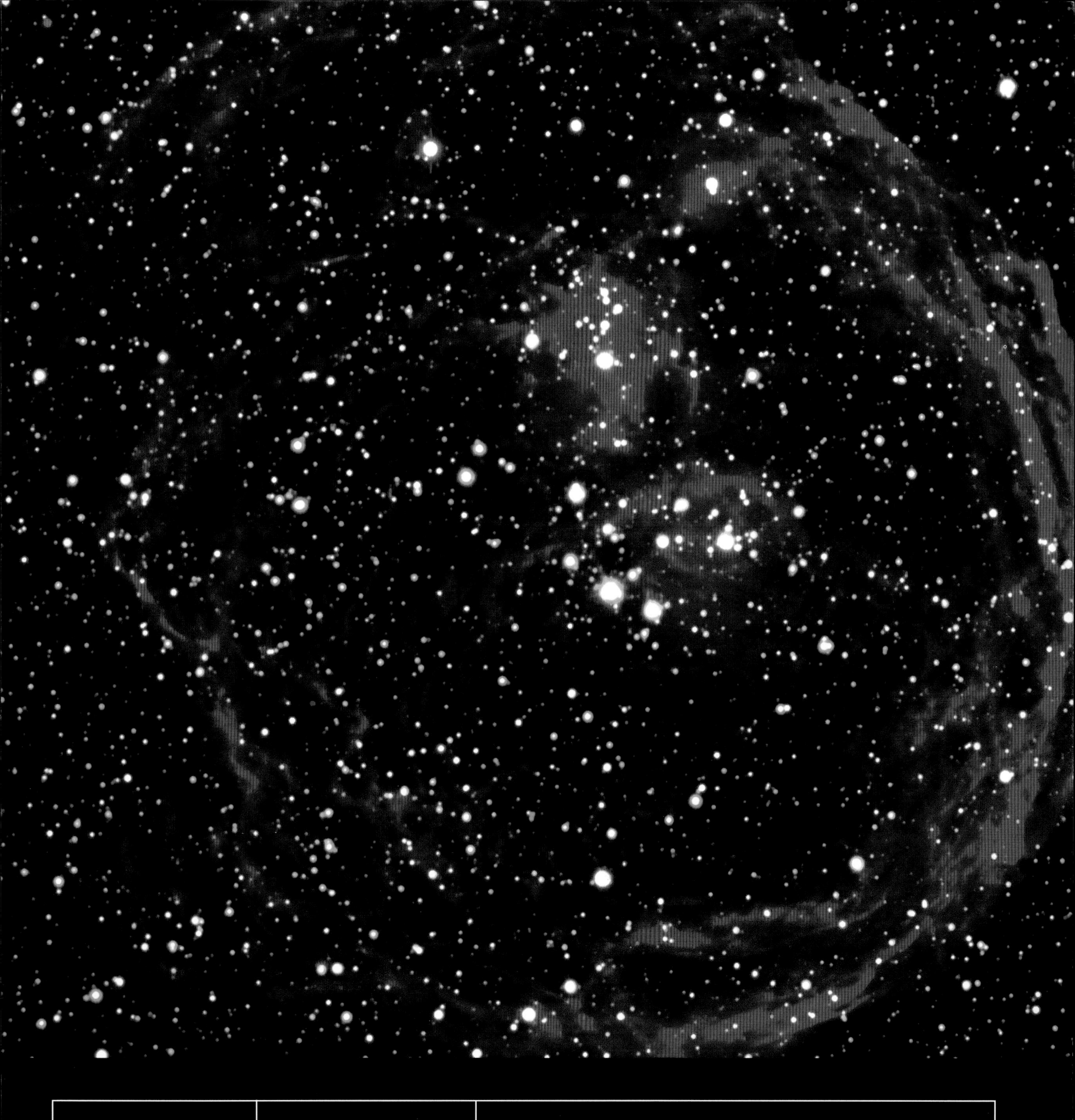

160 **thousand** light years	*N70* Emission nebula	This celestial chrysanthemum measures a prize-winning 300 light years in diameter. A brisk stellar wind from a cluster of stars behind the nebula creates a strong outflow of particles while the cluster's gravity attracts matter towards it. Where the outflowing and in-falling streams meet, gas is heated and the chrysanthemum blooms.

160 **thousand** light years	*Ghost Head Nebula* NGC 2080 Emission nebula	Looking into the eyes of this ghost we discover boiling clouds of hydrogen and oxygen superheated by newborn stars within. The left eye contains a single massive star, whereas the right eye is more complex and dusty, containing several stars. The stars of both eyes must be under 10,000 years old as their radiation has yet to burn off their natal gas shrouds.

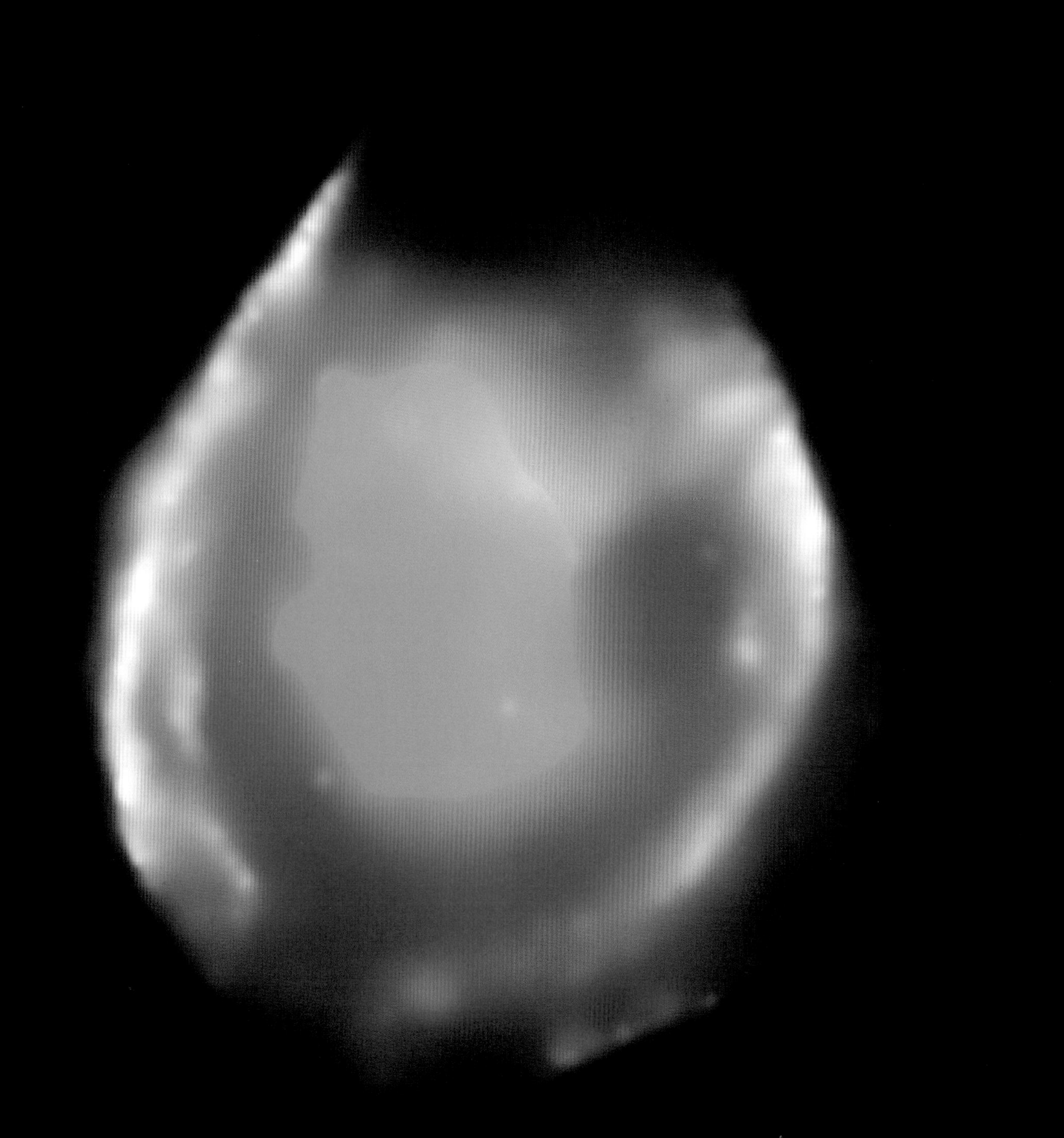

160 **thousand** light years	*DEM L71* Supernova remnant	Massive stars always end their fast and furious lives as supernovae, but it isn't an exclusive fate. This glowing x-ray shell reveals that low-mass stars can go out with a bang too – if the conditions are right. If a white dwarf can draw enough material from a companion star it will eventually collapse under its own weight triggering a thermonuclear explosion... as was the case here.

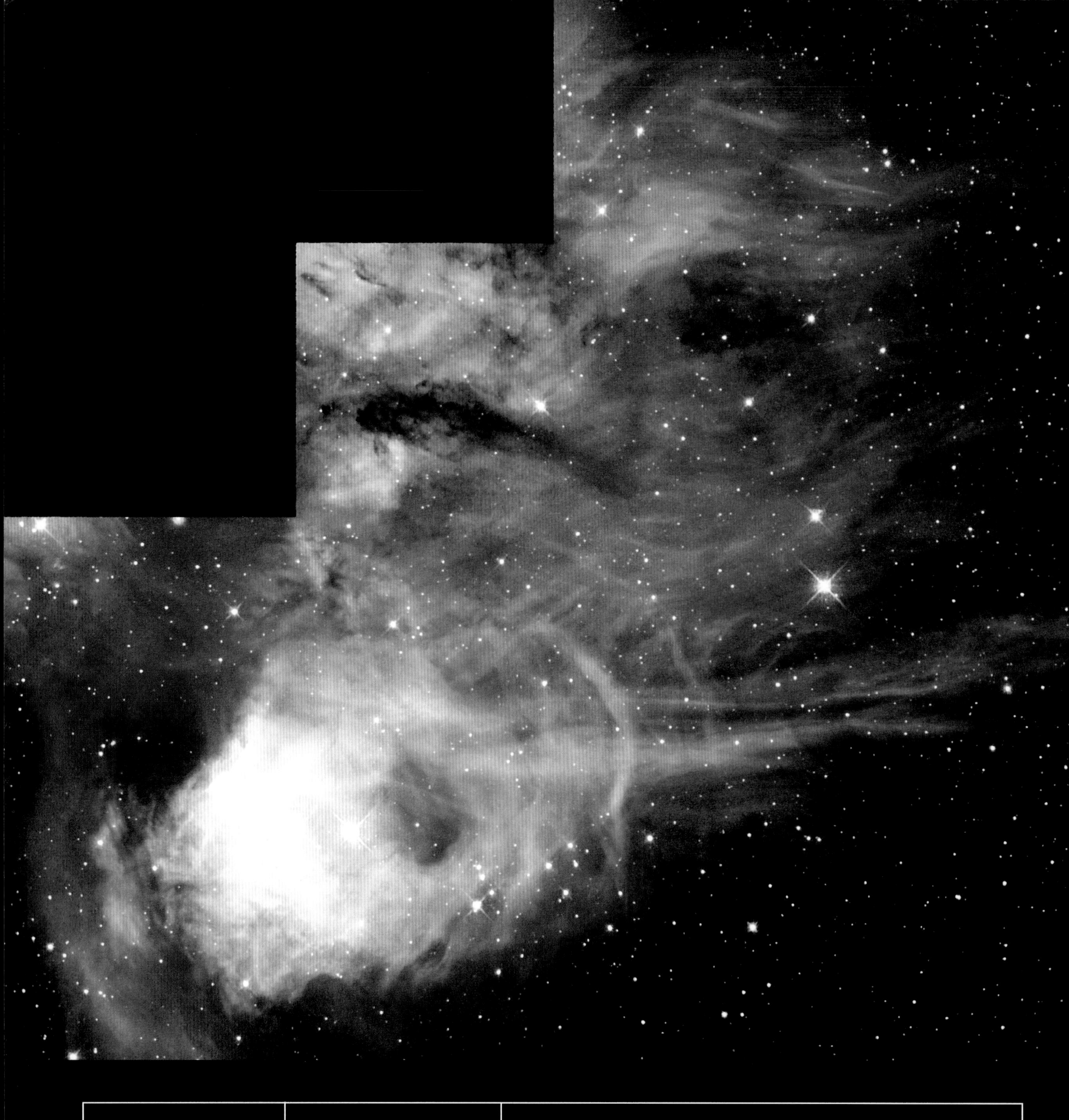

160 thousand light years	*N44C* Emission nebula	Gaseous filaments steam off this cauldron of a nebula. But why is it so hot? No-one is sure. The young star that powers N44C isn't hot enough on its own, and observation has revealed no other source, so some propose a companion neutron star, or even black hole, in a wide orbit. The missing heat could then be generated as it crashed through the nebula at the periapsis of its orbit.

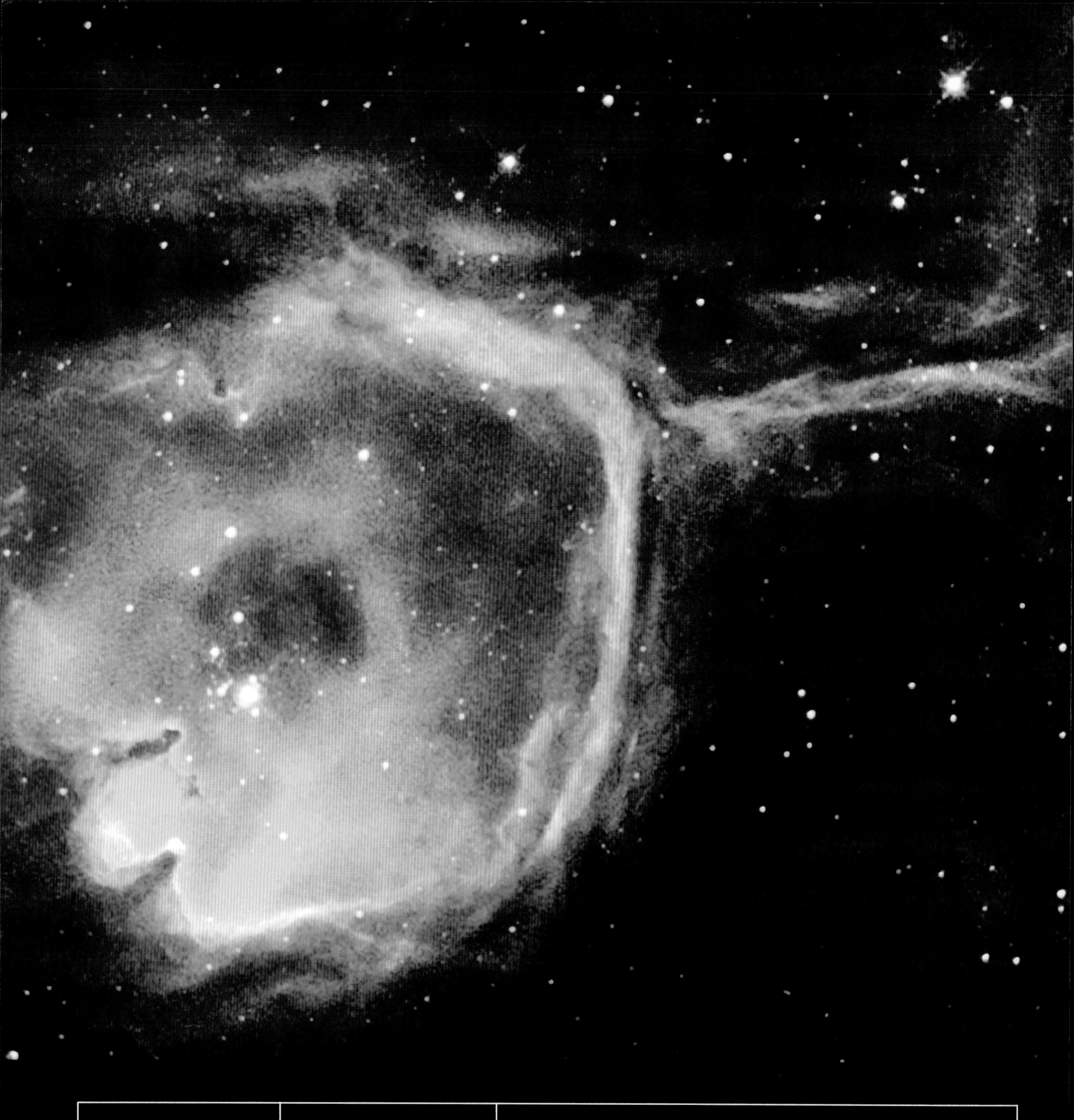

160 **thousand** light years	*N44F* Emission nebula	A celestial hurricane has inflated this bubble, 35 light years in diameter, from a dense cloud of dust and gas. Blowing at 6 million kph (4 million mph) this stellar wind emanates from the central blue star, which is throwing off more than 100 million times more mass per second than our Sun. It is this fast moving stream of particles that is pushing the surrounding clouds aside.

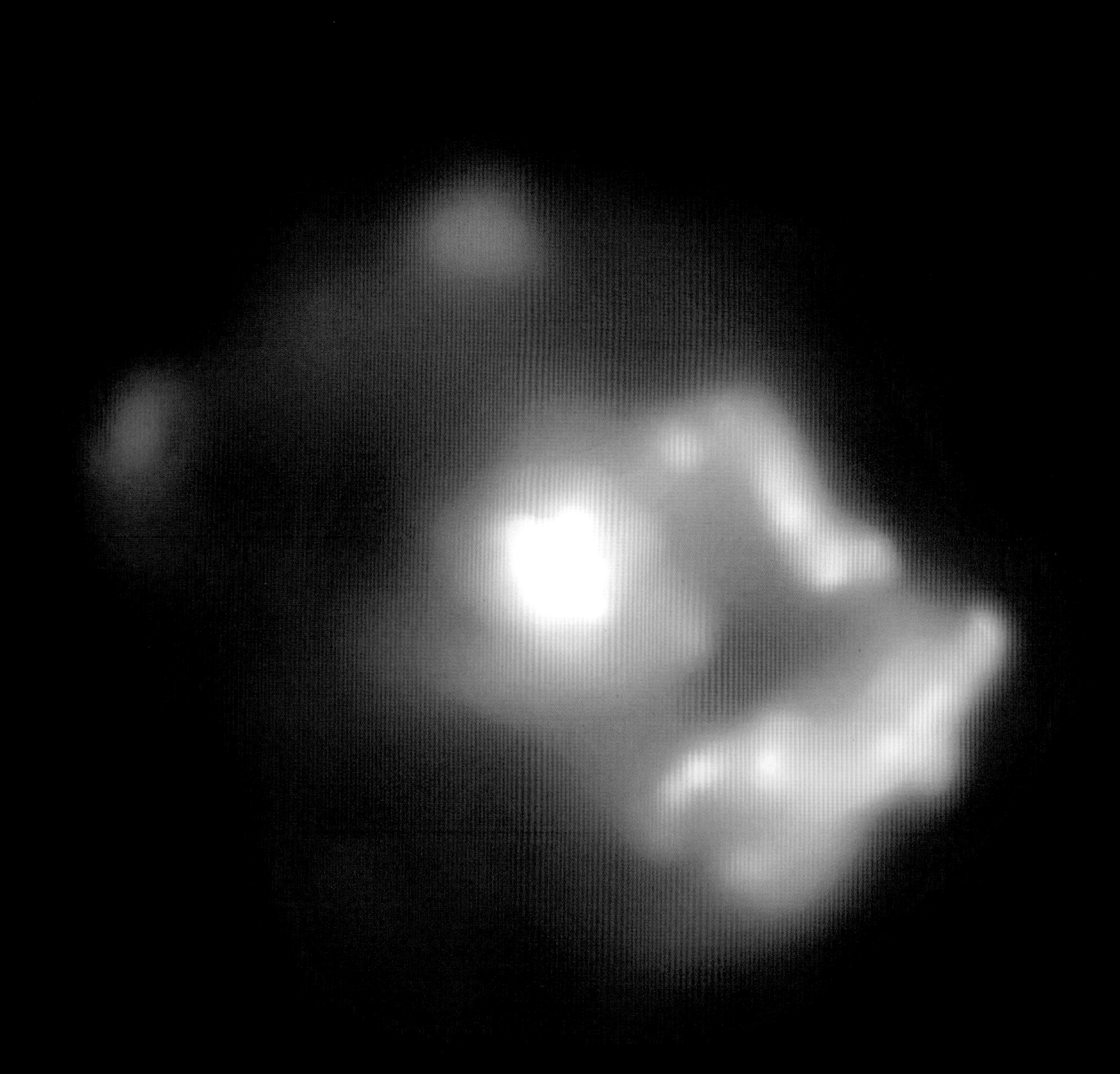

160 **thousand** light years	*SNR 0540-69.3* Pulsar	This ball of energy is a testament to the power unleashed by supernovae. At its core, embedded in a seething mass of high-energy particles, lies a pulsar spinning 20 times a second, generating power at a rate equivalent to 30,000 Suns. Surrounding this powerhouse is an envelope of gas, 40 light years in diameter, shock-heated to 50 million °C (90 million °F) by the passage of the original blast.

160 thousand light years	*NGC 1850* Star cluster	The Large Magellanic Cloud has revealed cosmic fauna with no known counterpart in our galaxy. These two star clusters look like globular clusters but are much younger. The main cluster is 50 million years old, while the smaller is only 4 million years old. These star factories have already produced some massive, short-lived stars, as the nebula of supernova debris on the left testifies.

160

thousand
light years

Henize 206

Emission nebula

Upon a 1,000 light year canvas, this portrait reveals a cycle of death and rebirth on a cosmic scale. Tell-tale sweeping arcs in the top half of the picture indicate that the shockwave from an ancient supernova has swept into the central cloud of hydrogen, compressing it and kick-starting a new generation of star formation. The death of one star leads to the birth of many.

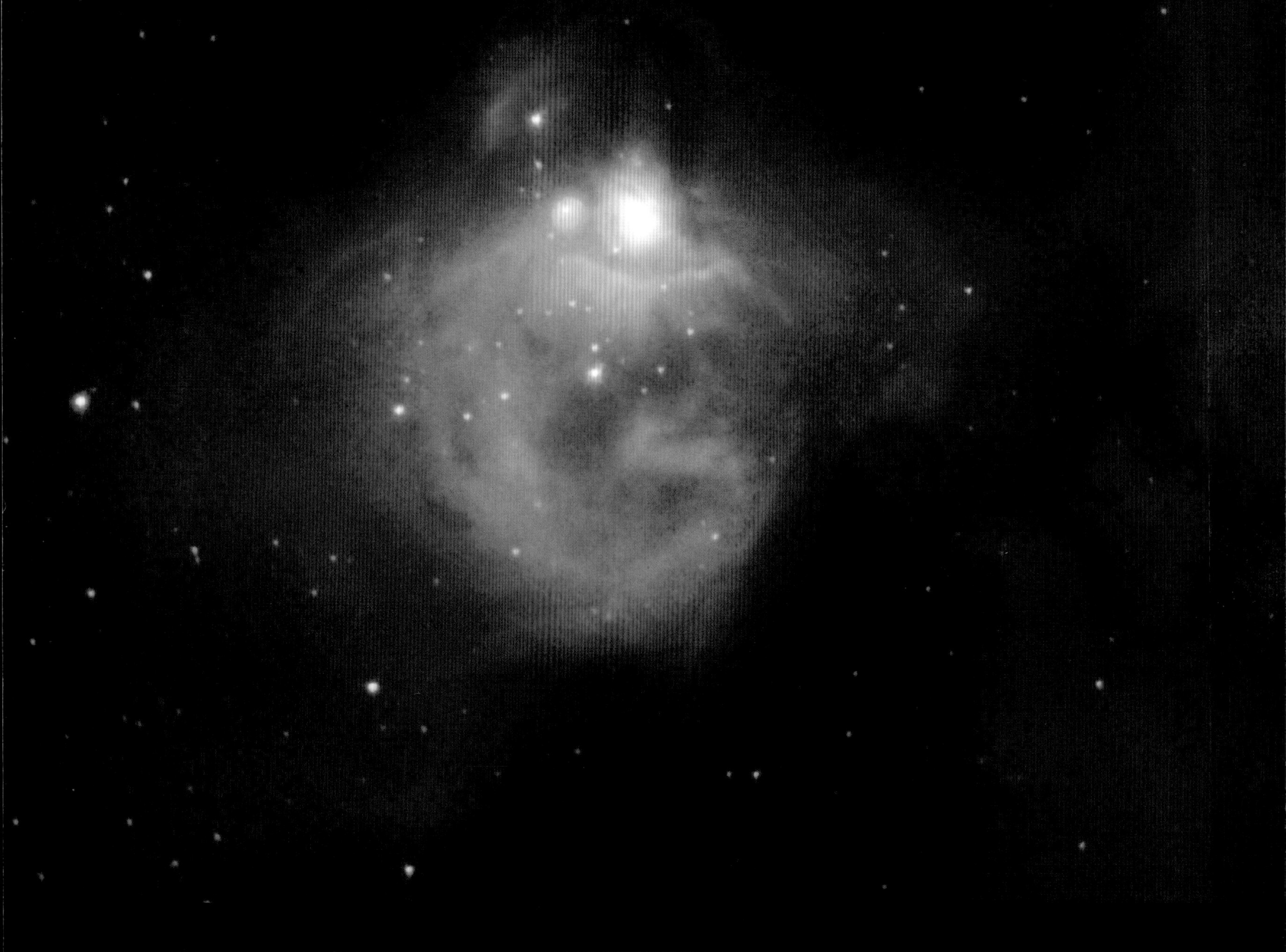

160 thousand light years	*NGC 1748* Emission nebula	This stellar cradle is being torn apart by radiation from the heavyweight infants it has birthed. The innocuous-looking star in the centre of the nebula is actually 30 times more massive than our Sun and nearly 200,000 times brighter – and it has triggered the birth of some even larger siblings: embedded in the bright region towards the top of NGC 1748 is a star with a mass of 45 Suns.

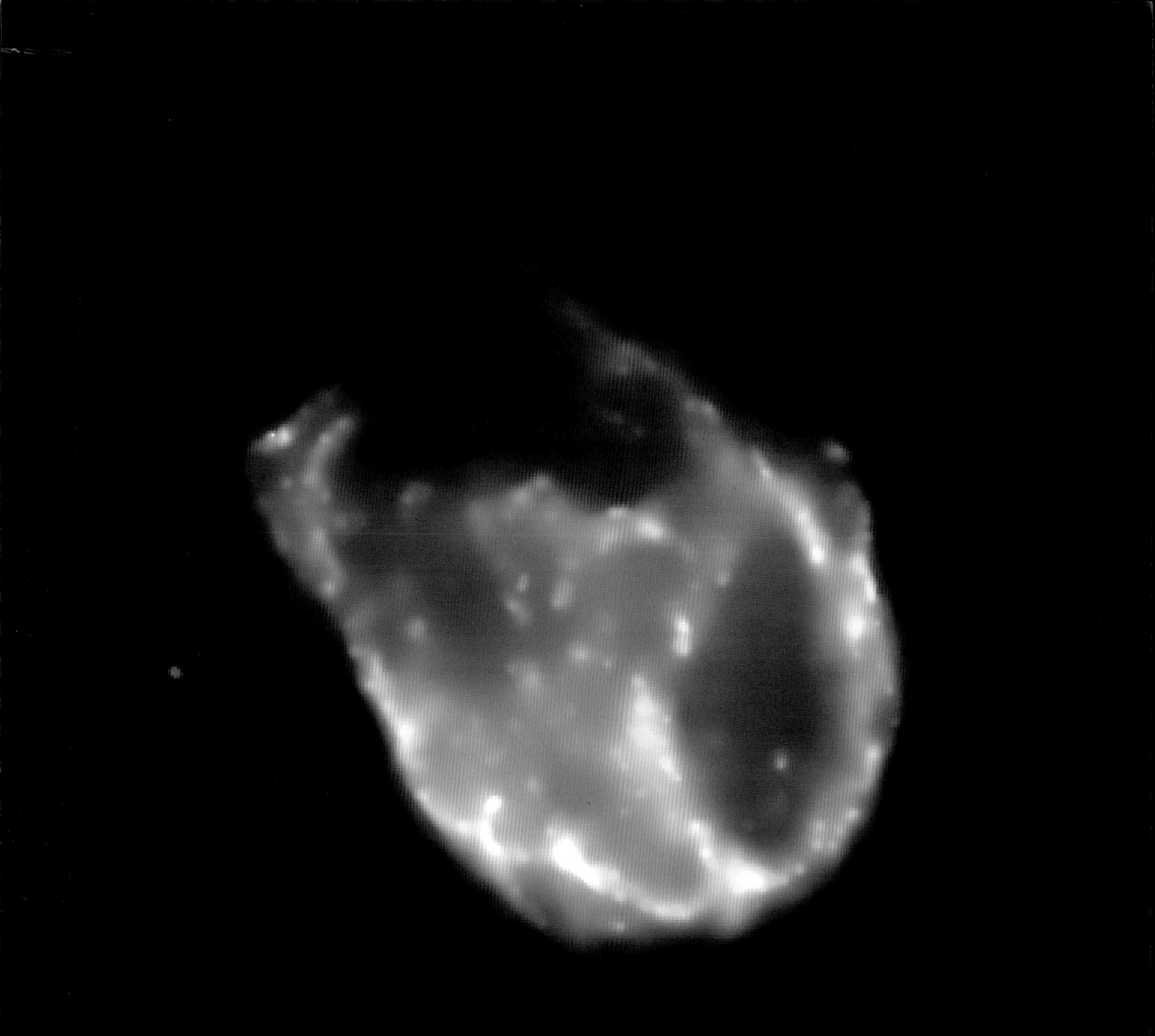

160 **thousand** light years	*N132D* Supernova remnant	A 10 million °C (18 million °F) supernova shockwave has collided with a giant molecular cloud, sweeping up material massing more than 600 Suns and lighting it up to create a phosphorescent jellyfish 80 light years across. The light from the original blast would have illuminated Earth's skies for a few weeks in approximately 1,000 BC.

170

thousand
light years

Hodge 301

Star cluster

This cluster of massive and aging stars (top right) lies within the Tarantula Nebula, the most fertile starburst region in the local universe. The largest of Hodge 301's inhabitants have ended their lives as supernovae, creating shockwaves that have ploughed into the surrounding nebula compressing the gas into a multitude of glowing sheets and filaments, as can be seen here.

170

thousand
light years

R136

Star cluster

Another denizen of the Tarantula Nebula is a cluster of the largest, hottest and most massive stars known. R136's blazing stars (above centre) are 10 times hotter than our Sun and up to 100 times more massive. Their prodigious energy has unleashed a wave of destruction and creation around them: collapsing the Tarantula's existing dust and gas clouds into incubators for nascent stars.

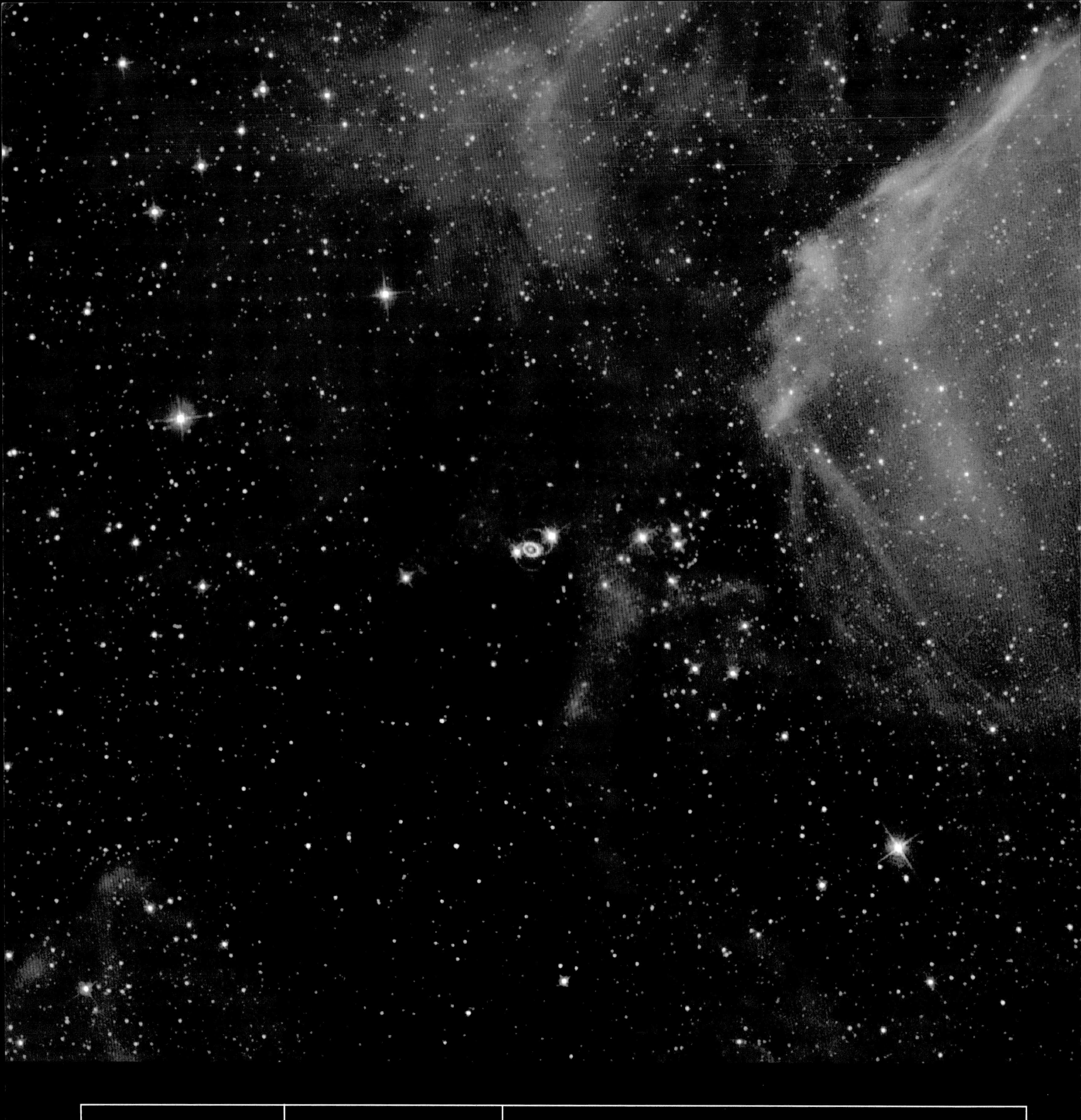

170 thousand light years	*SN1987a* Supernova remnant	The central figure of eight in this picture is the remnant of the closest supernova since the invention of the telescope. Spectacularly violent, SN1987a blazed with the power of 100 million Suns for several months in 1987. The surrounding gas clouds will soon be lit up by a 10 million °C (18 million °F) shockwave moving at 16 million kph (10 million mph). Watch this space.

200 **thousand** light years	*NGC 346* Emission nebula	Nebula NGC 346 in the Small Magellanic Cloud is home to a crowded nursery of infant stars a mere five million years old. These protostars will keep on growing, fed by the gravitational collapse of NGC 346's surrounding gas clouds, until they are massive enough to ignite the hydrogen they have gathered in a blaze of nuclear fusion.

200 thousand light years	*E0102-72* Supernova remnant	1,000 years ago in the Small Magellanic Cloud a star at least eight times more massive than the Sun exploded at speeds in excess of 20 million kph (12.5 million mph). Now forty light years across, these are its remains. At the hub of this cosmic wheel, as at the heart of most supernovae remnants, lies either a neutron star or, if the star was massive enough, a black hole.

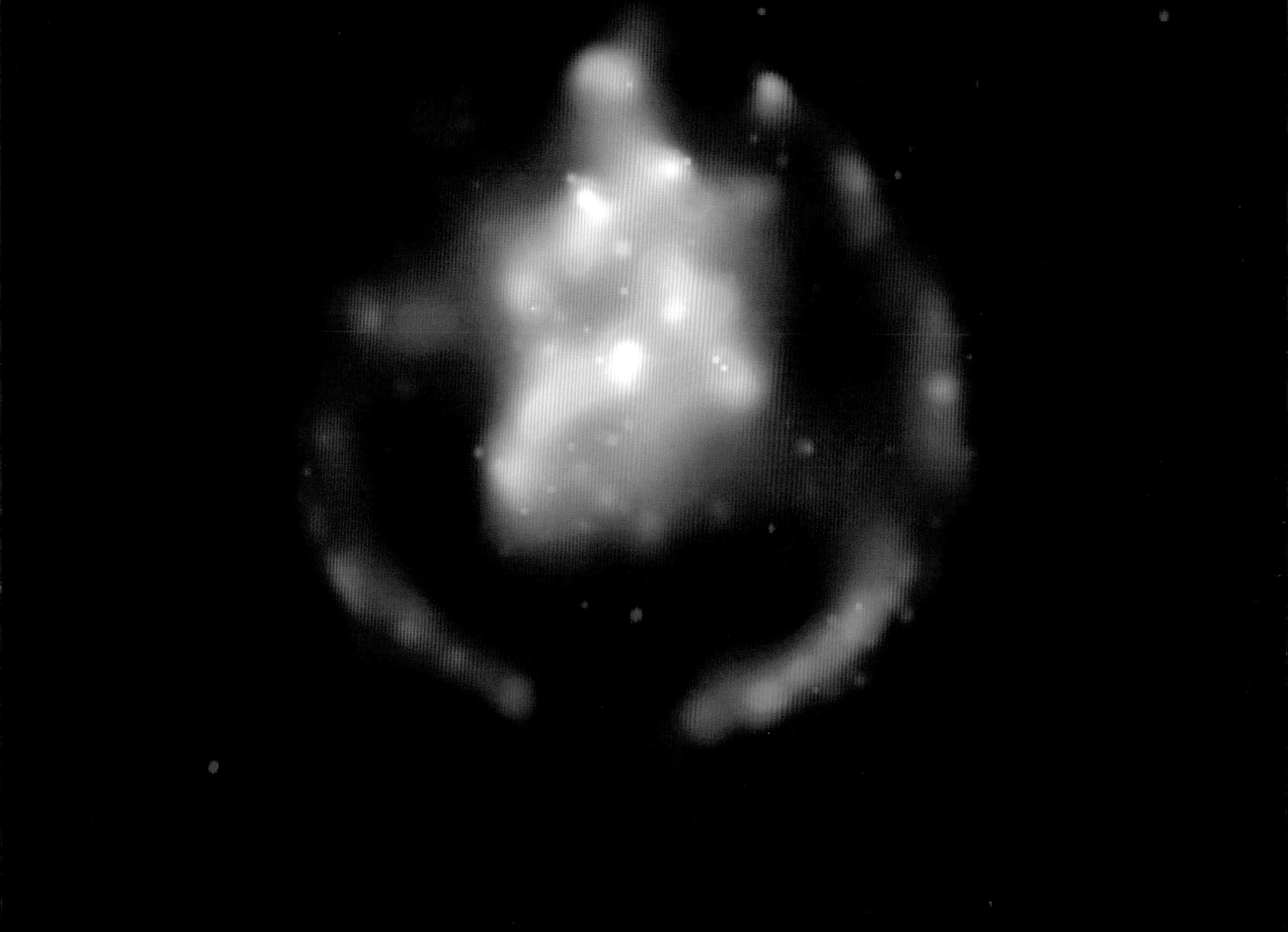

200

thousand
light years

SNR
0103-72.6

Supernova remnant

150 light years across, this shockwave of superheated gas is the remnant of another Small Magellanic Cloud supernova that exploded 10,000 years ago. Supernovae seed the universe with heavy elements forged in the stellar core for incorporation into future stars and planets. This remnant is particularly rich in oxygen and neon, meaning it was at least 10 times more massive than the Sun.

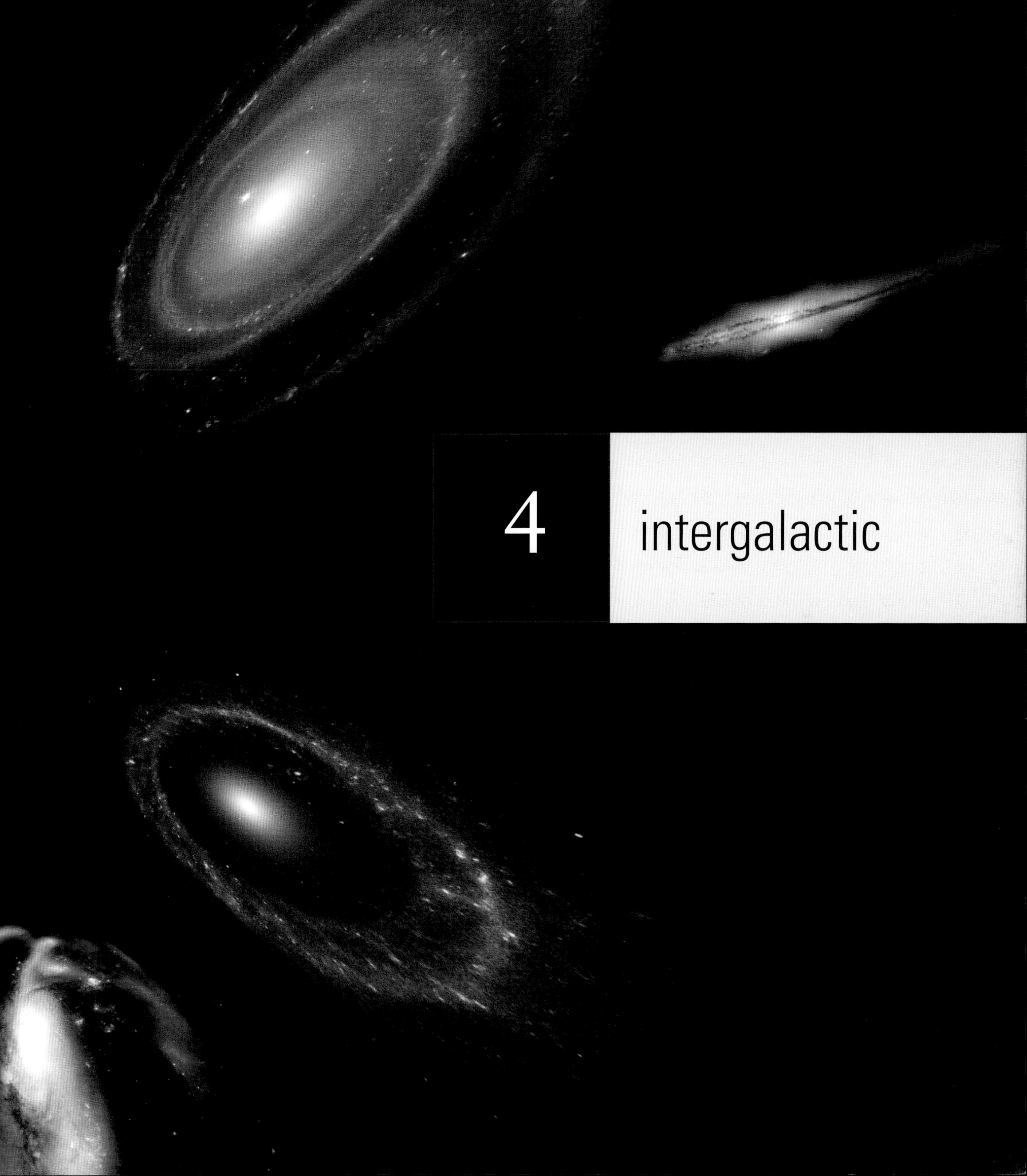

4 intergalactic

1.6 **million** light years	*Hubble-V* Emission nebula	The next step of our journey takes us deeper into the Local Group, to galaxies not in the immediate gravitational thrall of the Milky Way. Barnard's Galaxy (NGC 6822) is only a tenth the size of our galaxy but is remarkable for its prolific star forming clouds. One such cloud, Hubble-V, is aglow with the light of stars 100,000 times brighter than the Sun and measures over 200 light years across.

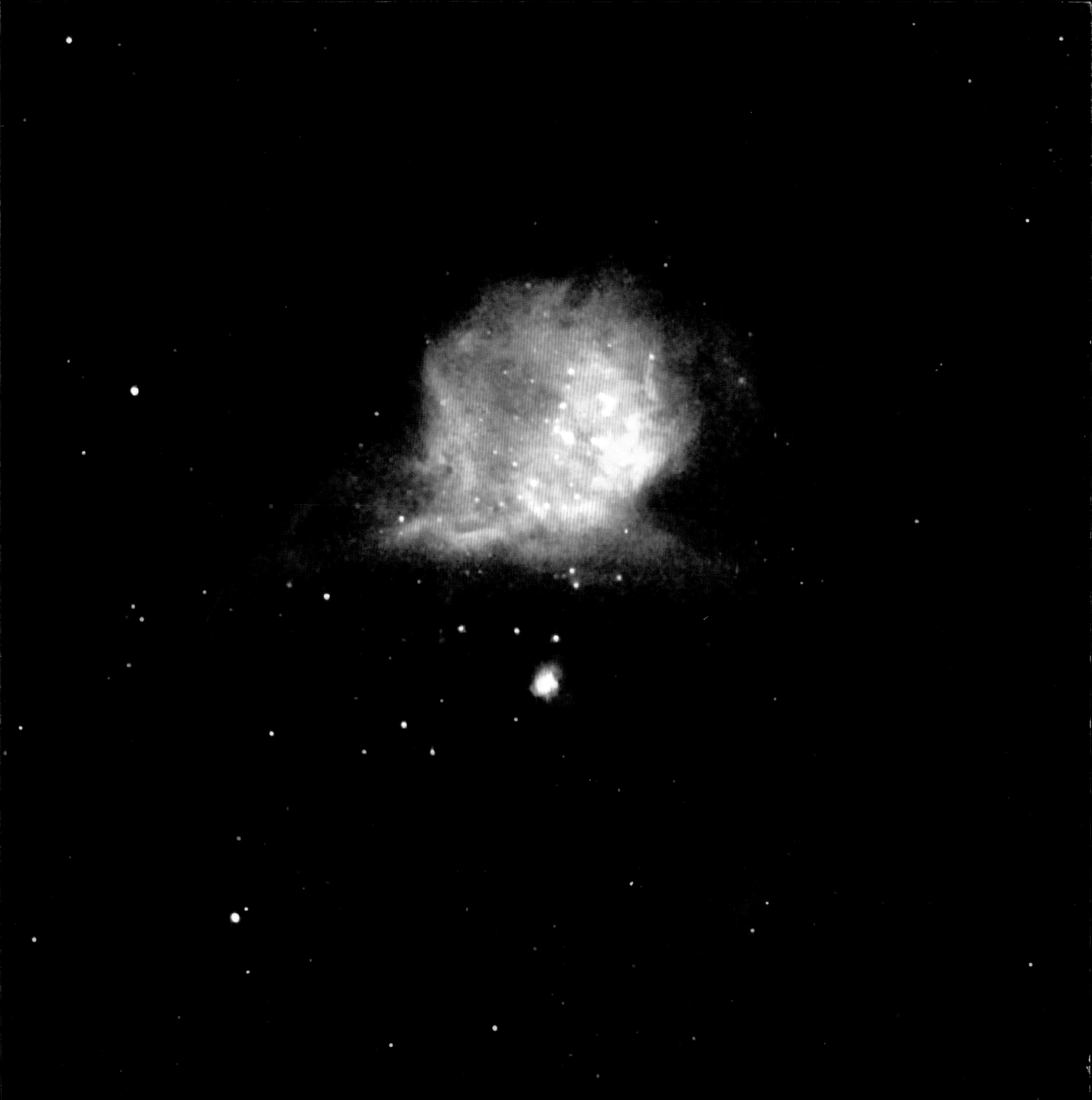

1.6 **million** light years	*Hubble-X* Emission nebula	Hubble-X is another of Barnard's Galaxy's stellar factories. At 100 light years in diameter it may be only half the size of Hubble-V but it is still over 10 times larger than our galaxy's Orion Nebula. With its huge gas clouds, ongoing evolution of massive star clusters, lack of heavy elements and irregular shape, Barnard's Galaxy is thought to replicate the universe's earliest galaxies.

2.5 **million** light years	*Andromeda* M31 Spiral galaxy	200,000 light years across and surrounded by a swarm of at least 10 satellite galaxies, Andromeda is the other titan of the Local Group. It may be twice as large as the Milky Way, but recent research suggests it is considerably less massive. As it is heading towards us at 500,000 kph (300,000 mph), in 3 billion years we will have a much more accurate estimate of its weight.

2.5 **million** light years	*Andromeda* M31 {detail} Spiral galaxy	Andromeda is no stranger to galactic collisions. A survey of 300,000 stars in its outer halo has revealed that one third of them are only 6 to 8 billion years old – just over half the age of the Milky Way's halo population. Whether these stars are the captured members of a smaller, younger galaxy or the progeny of a wave of star formation spurred by a larger collision is unknown.

2.7 **million** light years	*NGC 604* Star forming region	1,300 light years across and brewing stars over 100 times more massive than the Sun, a spiral arm of the Triangulum Galaxy (M33) holds the largest discrete cauldron of star formation in the Local Group – and the known universe. Laden with over 200 stellar hypergiants, only the Tarantula Nebula in the Large Magellanic Cloud can boast a greater fertility.

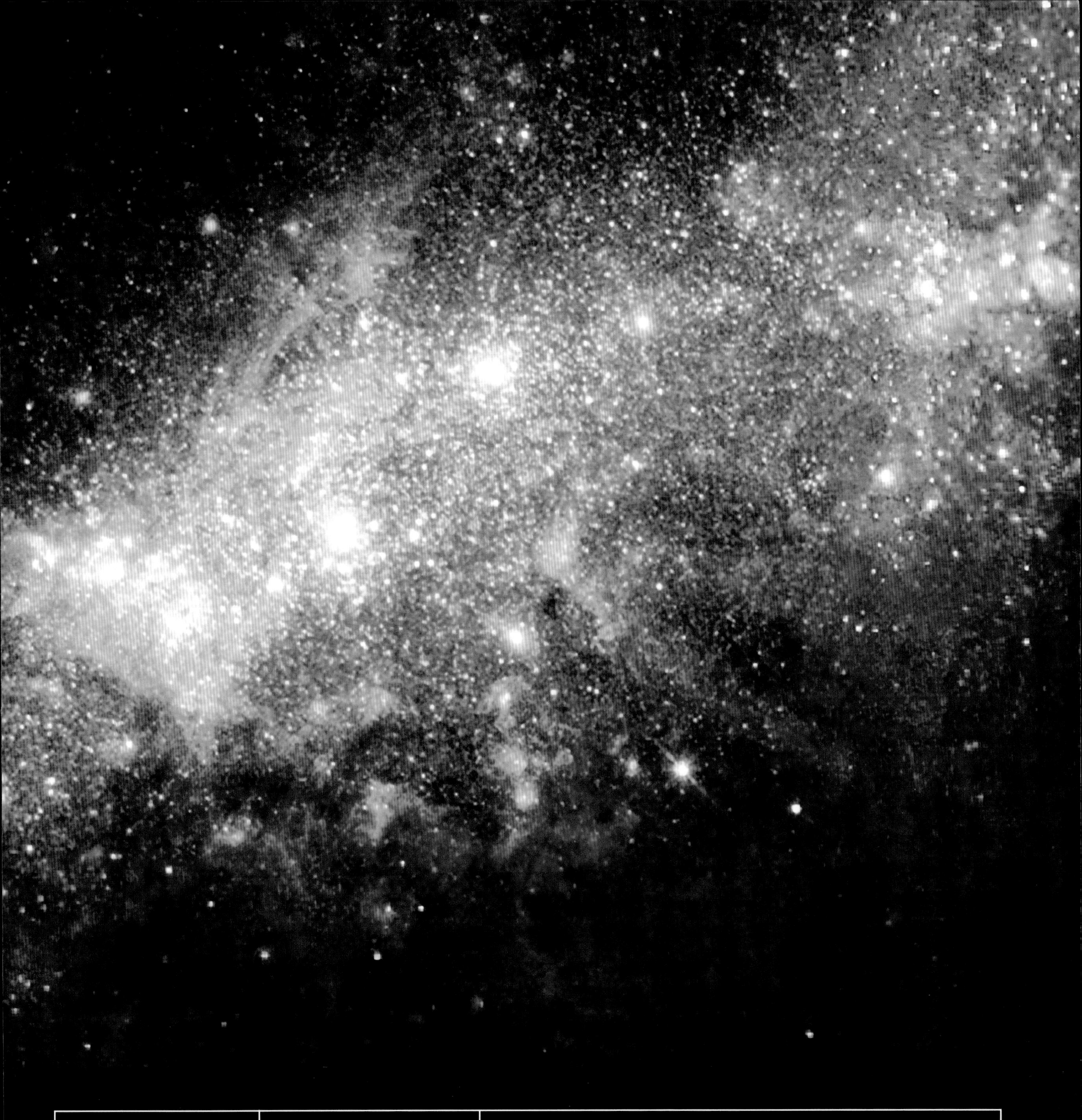

7

million
light years

NGC 1569

Starburst galaxy

Still recovering from a 20-million-year labour, the heart of dwarf galaxy NGC 1569 has given birth to two prominent star clusters. A close encounter or collision with another galaxy triggered a furious wave of stellar genesis that swept across NGC 1569 as a chain reaction of supernovae shocked the interstellar medium into successive generations of massive, short-lived stars.

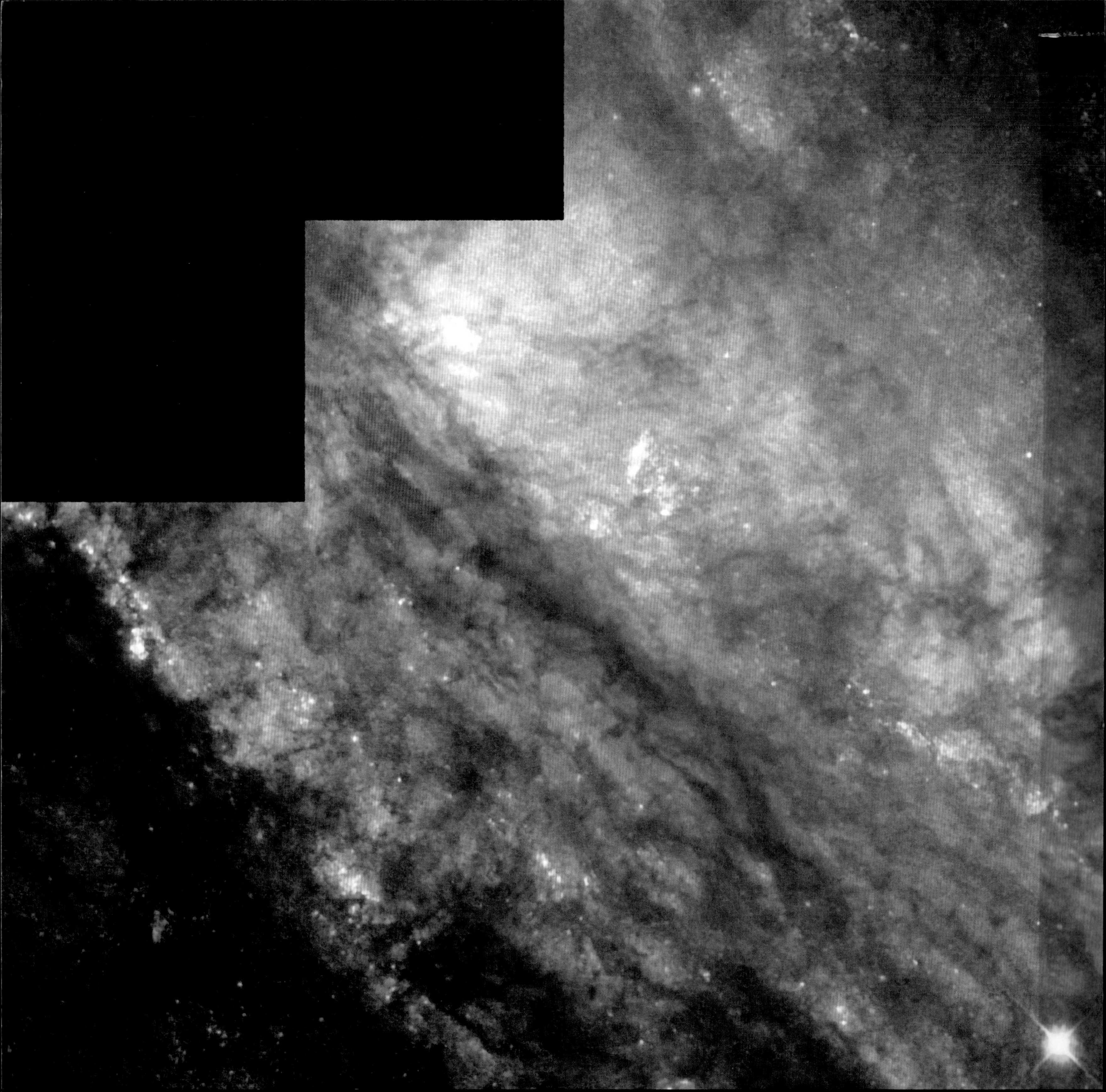

11 million light years	*Centaurus A* NGC 5128 Active galaxy	Like the old woman who swallowed a spider, Centaurus A has an entire spiral galaxy wriggling and jiggling and tickling inside her. Devoured over 100 million years ago, this supersized meal has fuelled a maelstrom of star formation and feeds a billion-solar-mass black hole – the engine that propels twin jets of radiation 25,000 light years into space and gives this galaxy its 'active' label.

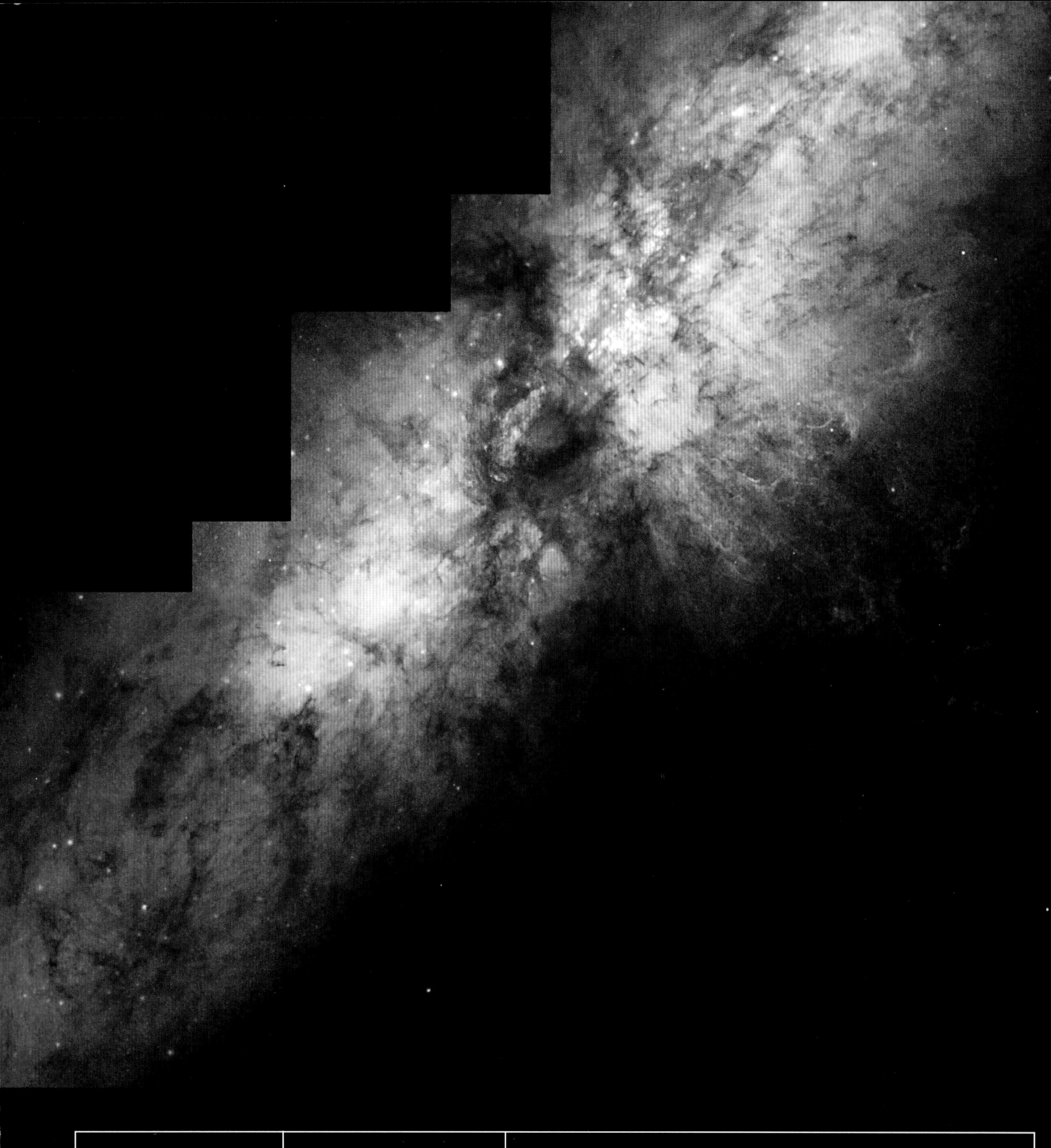

12 **million** light years	*M82* Starburst galaxy	600 million years ago, a close encounter with M81 ignited M82, an otherwise quiescent spiral galaxy, in a blaze of stellar creation. Wracked by gravitational tides, vast clouds of dust and gas around the galactic core collapsed, forming over 100 new globular clusters of more than 100,000 stars. Star birth continues today as debris scooped out by the near miss rains back down on M82.

12

million
light years

M81

Spiral galaxy

Spiral galaxies owe their elaborate anatomy to an ephemeral light show that traces the progress of a density wave churning through their discs. The crest of the wave is illuminated by a surf of massive, short-lived stars sculpted from the interstellar medium by the shock of its passage, as well as by existing stars bunching together as they slide over its peak.

13
million
light years

Circinus Galaxy ESO 97-G13

Active galaxy

The heart of the Circinus Galaxy glows vividly as clouds of gas spiral into a supermassive black hole. The plunge towards the event horizon friction heats the clouds to a temperature of many millions of degrees, creating a boiling wind that tears gas from the black hole's grasp and ejects it into space; seen here as the magenta streamers fluttering some 3,000 light years above the galactic core.

13 million light years	*NGC 4214* Starburst galaxy	Dense clouds of dust and gas fuel the starburst conflagration that consumes this irregular galaxy. Clusters of hypergiants so massive that their lifetime is measured in mere millions of years bloom only briefly but sow the seeds of the next generation. Their furious radiation and inevitable final detonation propagate shockwaves through the interstellar medium, forcing its collapse into new stars.

17 **million** light years	*Black Eye Galaxy* M64 Spiral galaxy	Behind a thick mask of dust, this galactic cannibal has been digesting its last meal for a billion years. Utterly consumed, there is no trace of the victim bar an outlying halo of gas that still rotates in the opposite direction to the rest of the galaxy. Where this disc rubs against M64 it has kindled a rash of hot blue stars surrounded by glowing red clouds of hydrogen.

24 **million** light years	*NGC 2787* Lenticular galaxy	Lenticular galaxies are believed to be spiral galaxies that have lost their limbs. The wave of star birth that forms the sweep of a spiral galaxy's arms ultimately drains the interstellar medium of its raw materials. With no more newborn stars to light their progress, the spirals simply evaporate. The star-like points of light orbiting NGC 2787's core are actually ancient globular clusters.

28
million
light years

Sombrero Galaxy M104

Spiral galaxy

55,000 light years across, with a mass equivalent to 800 billion stars, this huge galaxy gets its name from its unusually brilliant and bulbous core (containing over 2,000 globular clusters – 10 times as many as orbit our galaxy) and the dark gas lanes seen edge-on in its disc. X-ray emissions suggest that at its heart the Sombrero Galaxy harbours a black hole with the mass of a billion Suns.

million
light years

Whirlpool Galaxy NGC 5194

Spiral galaxy

The Whirlpool Galaxy is currently dismembering its dwarf companion, NGC 5195, but not without suffering a wave of gravitational remodelling itself. The stress of the encounter is responsible for the tightly wound spirals that give the galaxy its distinctive appearance, igniting sweeping gunpowder trails of star birth accentuated by the red flare of ionized hydrogen clouds.

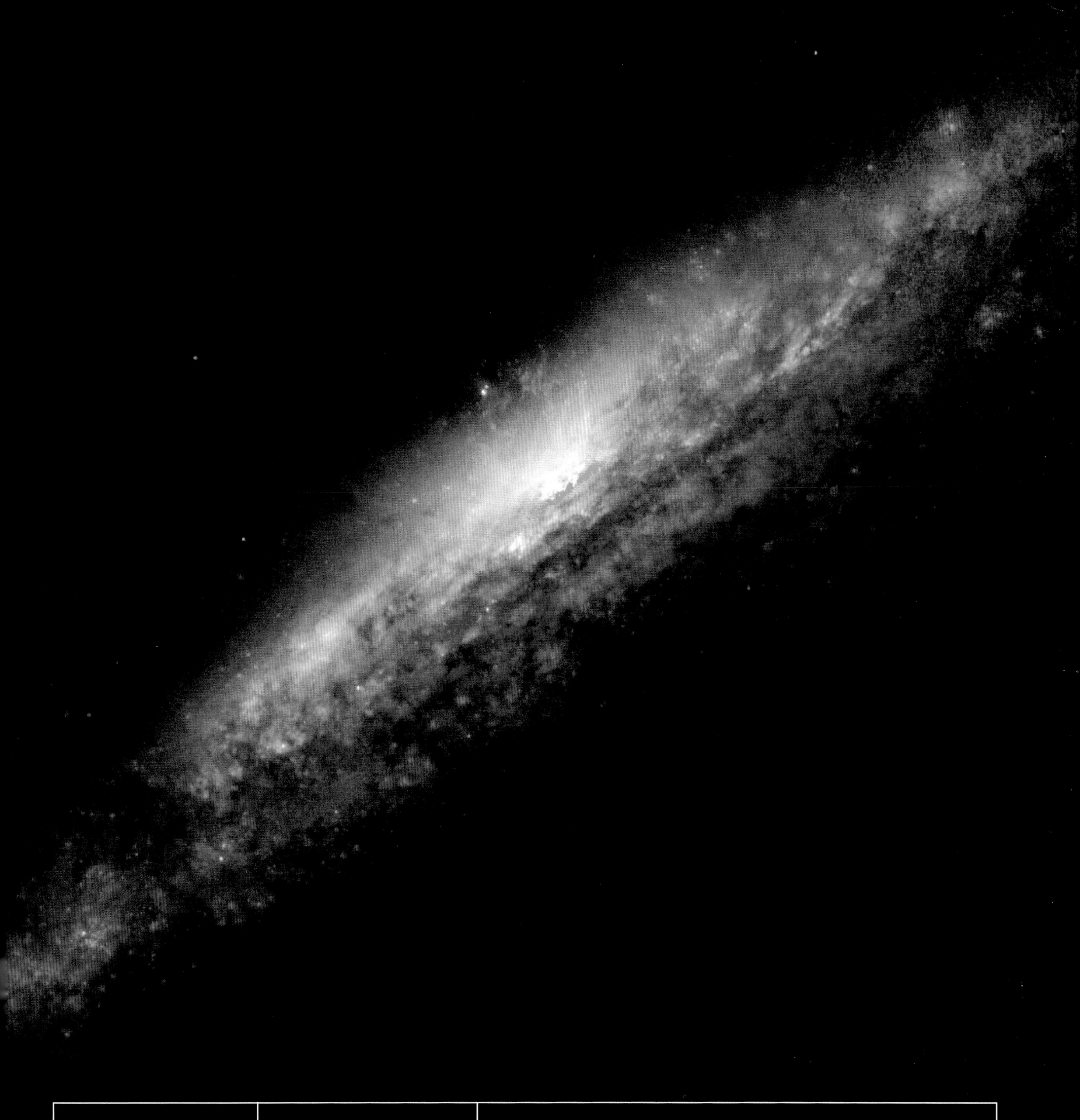

50 **million** light years	*NGC 3079* Spiral galaxy	Driven before a 6 million kph (4 million mph) gale of high-energy particles, fountains of 10 million °C (18 million °F) gas tower 3,500 light years above NGC 3079's core. This superwind blows either from the accretion disc of a super-massive black hole or from a titanic outbreak of starburst activity. Eventually, the gas will rain back down on the galaxy seeding a new generation of stars.

50 **million** light years	*M87* NGC 4486 Active galaxy	Dominating the Virgo Cluster of galaxies, M87 is a giant elliptical galaxy 20 times more massive than the Milky Way. A population of over four trillion stars provides a rich diet for the two-billion-solar-mass black hole lurking in its core. As dust, gas and entire suns spiral into its maw, a fusillade of radiation is unleashed, piercing over 5,000 light years of intergalactic space.

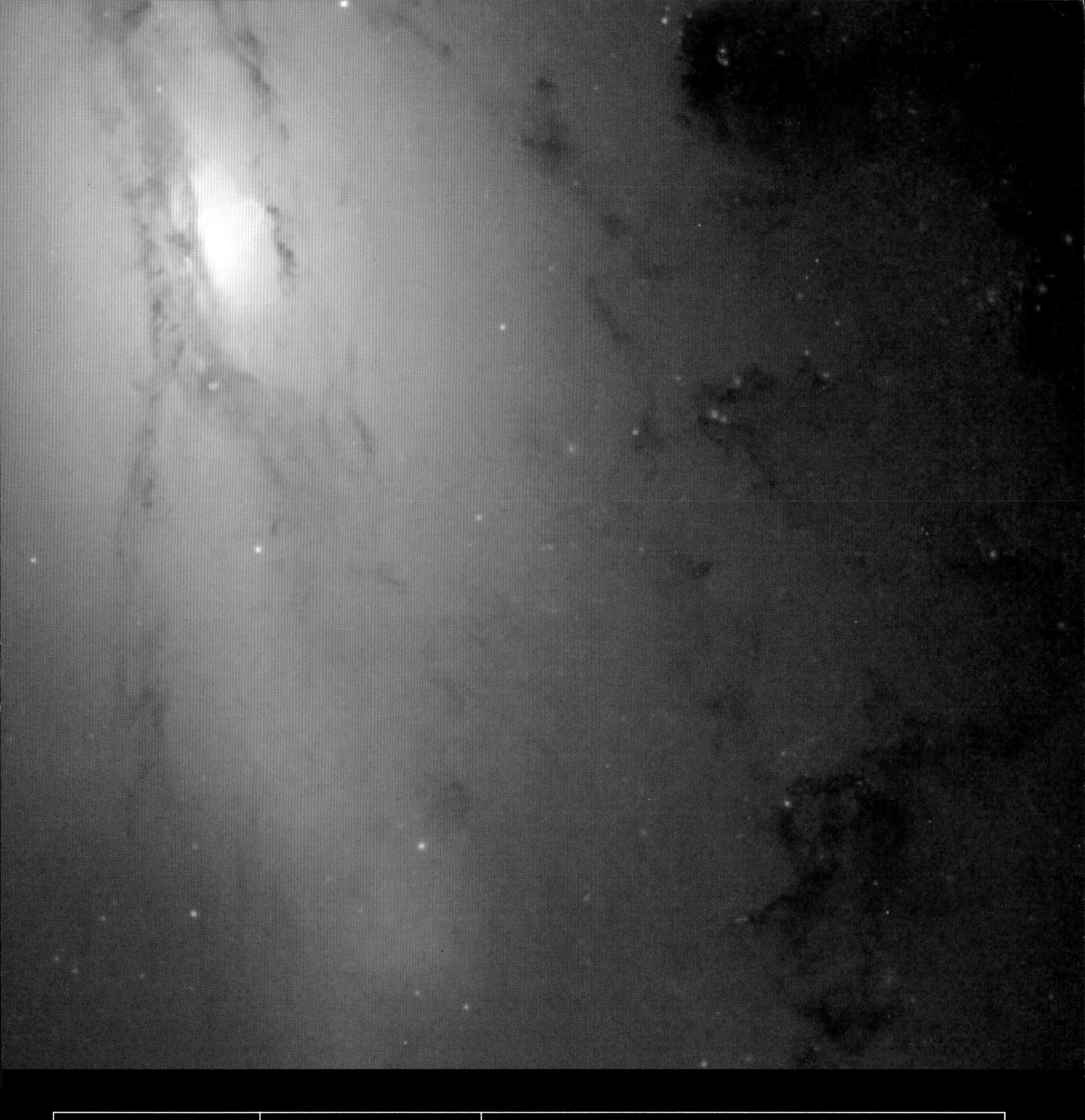

50

million
light years

NGC 4438

Active galaxy

NGC 4438 hosts a black hole that has inflated an 800-light-year bloom of plasma from its accretion disc (top left). Such potent displays are only possible because black holes are the most efficient engines in the universe: as material falls towards the event horizon up to 50 percent of its mass can be directly converted into energy. As a comparison, nuclear fusion is less than one percent efficient.

55 **million** light years	*NGC 4013* Spiral galaxy	Dark filaments of dust have formed a reef 500 light years deep, neatly bisecting this edge-on spiral galaxy. Containing the raw materials for many millions of suns, the reef is a stellar spawning ground and sparkles with the ultraviolet radiation of newborn clusters of hypergiant stars . The bright object near the core is neither a supernova nor an active black hole – merely a star in our galaxy.

60 **million** light years	*NGC 4414* Spiral galaxy	Lacking well-defined spiral arms, the woolly appearance of NGC 4414 classifies it as a flocculent spiral galaxy. Computer simulations suggest such spirals are conjured not from the galactic sweep of density waves, but self-propagate as cycles of stellar genesis and apocalypse stimulate similar episodes in adjacent regions. Galactic rotation then weaves these starbursts into a spiral pattern.

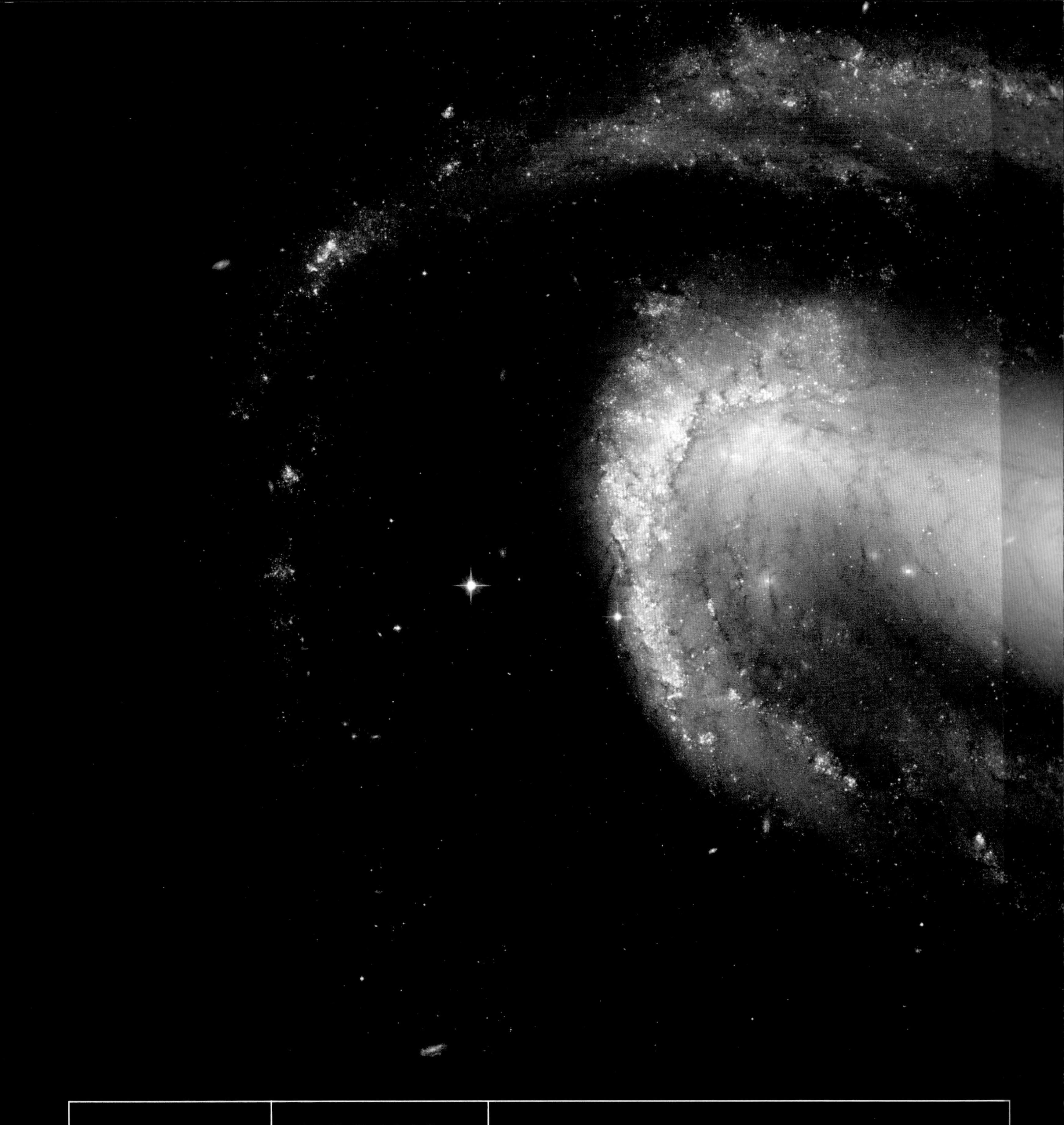

63 **million** light years	*NGC 1300* Spiral galaxy	Our tour of galactic archetypes continues with this barred spiral galaxy. Rather than springing directly from the galactic nucleus, this galaxy's arms attach to a bar of dust, gas and stars extending 15,000 light years either side of it. These bars are believed to funnel material towards the core, fuelling star formation or alternatively feeding a supermassive black hole.

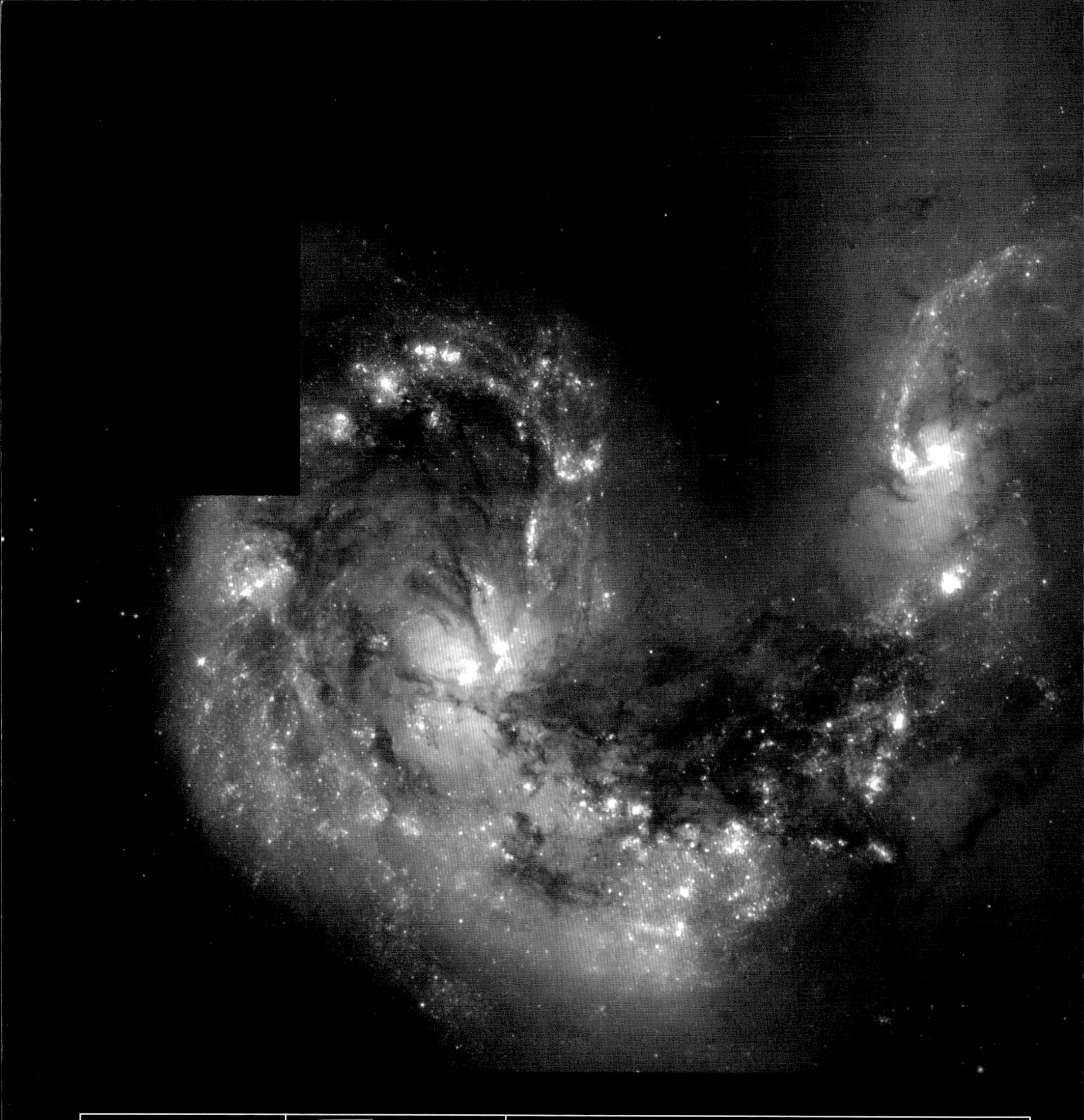

63 **million** light years	*Antennae Galaxies* NGC 4038 / 9	The collision of two galaxies has engulfed both bodies in a billion-year spree of stellar construction. Brilliant knots of light herald the creation of thousands of star clusters bursting with hypergiant suns. Such massive stars burn brightly but only briefly: x-ray surveys have revealed both galaxies to be strewn with the collapsar ashes of their predecessors.

98

million
light years

NGC 3370

Spiral galaxy

Now serene against a canvas of more distant galaxies, in November 1994 all the stars of NGC 3370 were briefly outshone by the detonation of a single white dwarf (a Type 1a supernova). Of course, with the stellar population of the observable universe estimated at over a billion trillion, supernovae aren't a rare phenomenon and occur somewhere in the universe every few seconds.

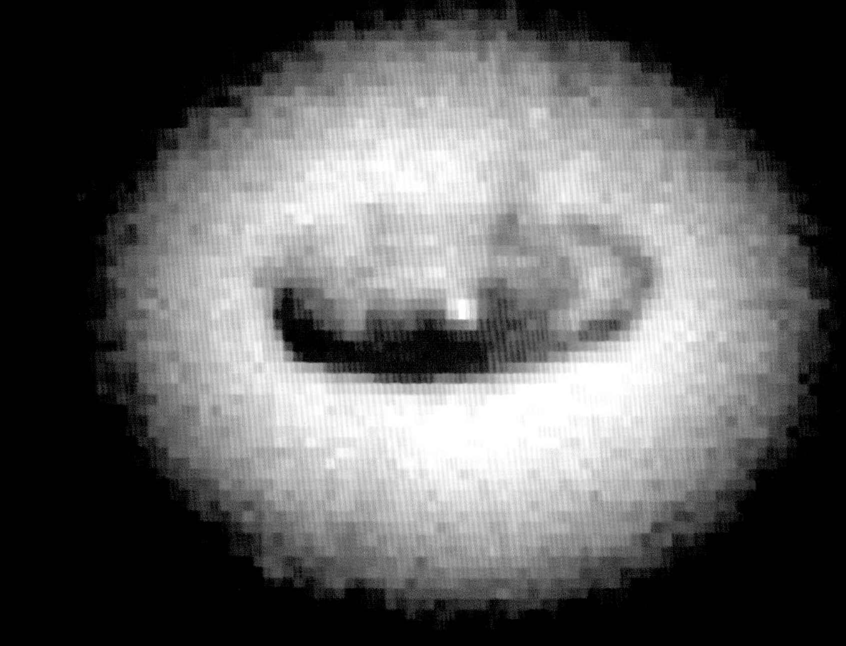

100 **million** light years	*NGC 4261* {detail} Black hole	Cloaked in an accretion disc 800 light years across, a black hole larger than our solar system stalks the core of NGC 4261. Weighing in at 1.2 billion solar masses it has grown fat on the remnants of a small galaxy captured by NGC 4261. An honour guard of collapsars attend this leviathan – the legacy of a wave of star formation stirred by its victim's dismemberment.

111
million
light years

NGC 4622

Spiral galaxy

NGC 4622 is a unique specimen among spiral galaxies, as it appears to be spinning backwards. From the orientation of its outer spiral arms one would expect a counterclockwise rotation, but observation reveals that it is, in fact, turning in the opposite direction. As with most peculiar galaxies, a galactic merger is thought to be to blame.

114

million
light years

NGC 2207 & IC 2163

Spiral galaxies

Locked in a violent tarantella, IC 2136 unravels as it swings round its larger partner, flinging a trailing arm of stars 100,000 light years across space. Their dance will continue for billions of years before IC 2163 finally surrenders to NGC 2207's embrace. In the crowded early universe such galactic mergers were the rule rather than the exception.

115

million
light years

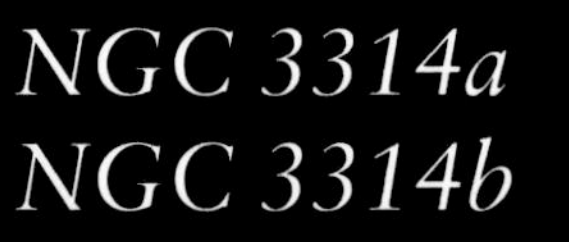

NGC 3314a
NGC 3314b

Spiral galaxies

A chance alignment captures two galaxies slipping past each other like ships passing in the night. NGC 3314a (foreground) has steered a safe course 25 million light years clear of NGC 3314b. Silhouetted against the glow of the background galaxy, NGC 3314a reveals a disc stippled by dark lanes of dust that echo its luminous star-laden limbs.

150

million
light years

ESO 510-G13

Spiral galaxy

The echo of a galactic collision is recorded in the warp of ESO 510-G13's disc. Despite the obvious violence of the impact, actual stellar collisions will have been vanishingly rare – the distances between stars in any galaxy are simply too vast, even in the densest clusters. Interstellar clouds of dust and gas however, cannot avoid being smashed together and are fated to collapse into new stars.

235

million
light years

Perseus A

NGC 1275

Active galaxy

The heart of the Perseus Cluster resounds (quite literally as we shall discover) with the clash of two galaxies. At 10 million kph (6 million mph), a sprial galaxy has scythed into an elliptical galaxy, preciptating an eruption of new star clusters. The impact has also woken a supermassive black hole, which showers Perseus with a torrent of radiation as it feeds on a surfeit of displaced material.

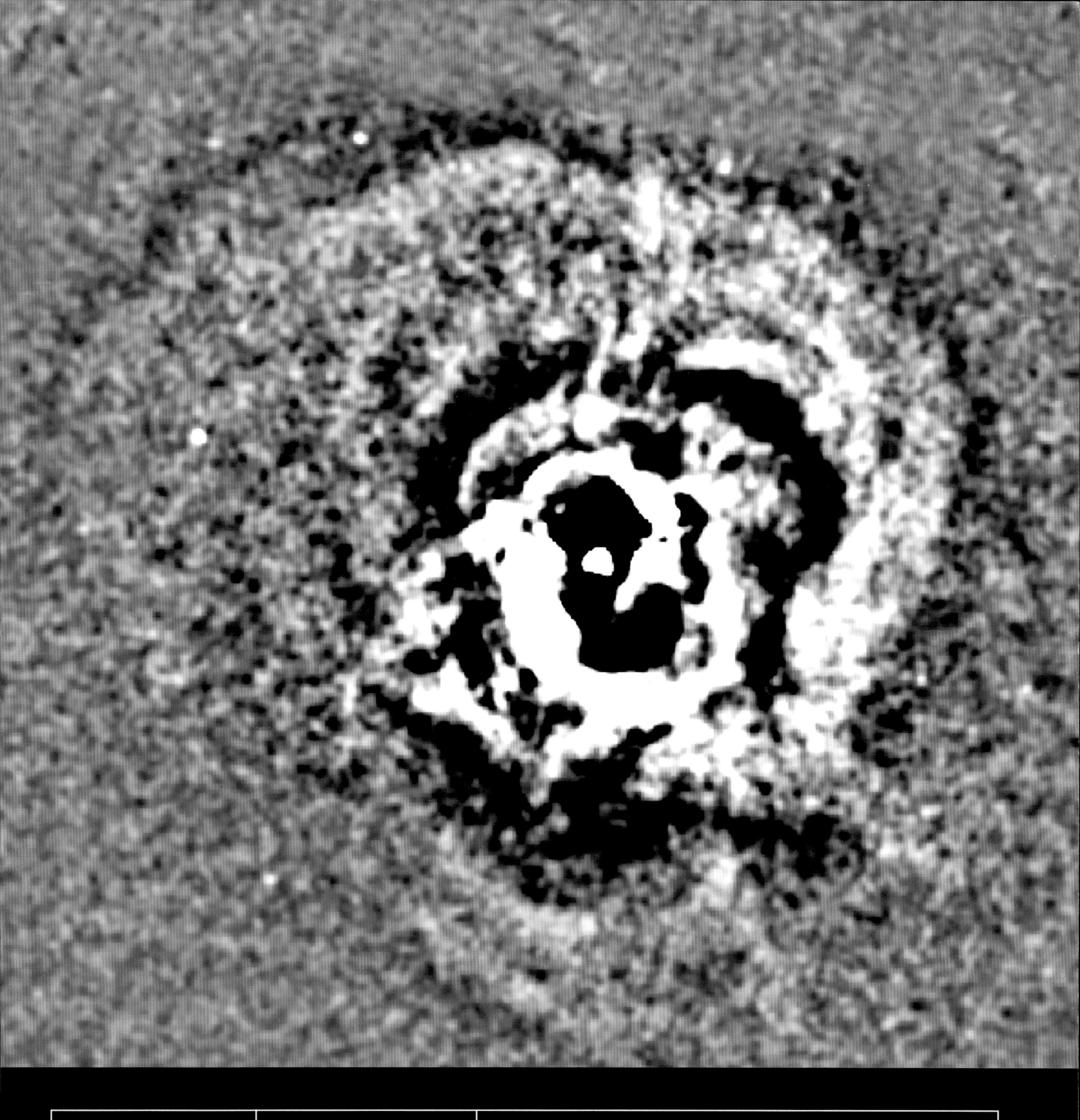

235

million
light years

Perseus A

NGC 1275 {detail}

Black hole

In space no one can hear you scream – unless you are a supermassive black hole. Ripples in the swirl of gas being sucked into the dark heart of Perseus A reveal the presence of soundwaves. So what song sings this stellar gorgon for its supper? A simple, if guttural, melody: B-flat, fifty seven octaves beneath middle-C and a million billion times deeper than the limits of human hearing.

270
million
light years

Stephan's Quintet

NGC 7318a, b, 7319, 7320

We move from a single galactic collision in Perseus to a multiple pile-up in Pegasus: on the left, NGC 7318b barrels past 7318a, stoking starbursts in both galaxies; to the lower right 7319 trails a wounded arm from an encounter with the errant fifth member of the group, 7320c, now out of shot and half a million light years away; sliding off the top of the image, 7320 has so far escaped injury.

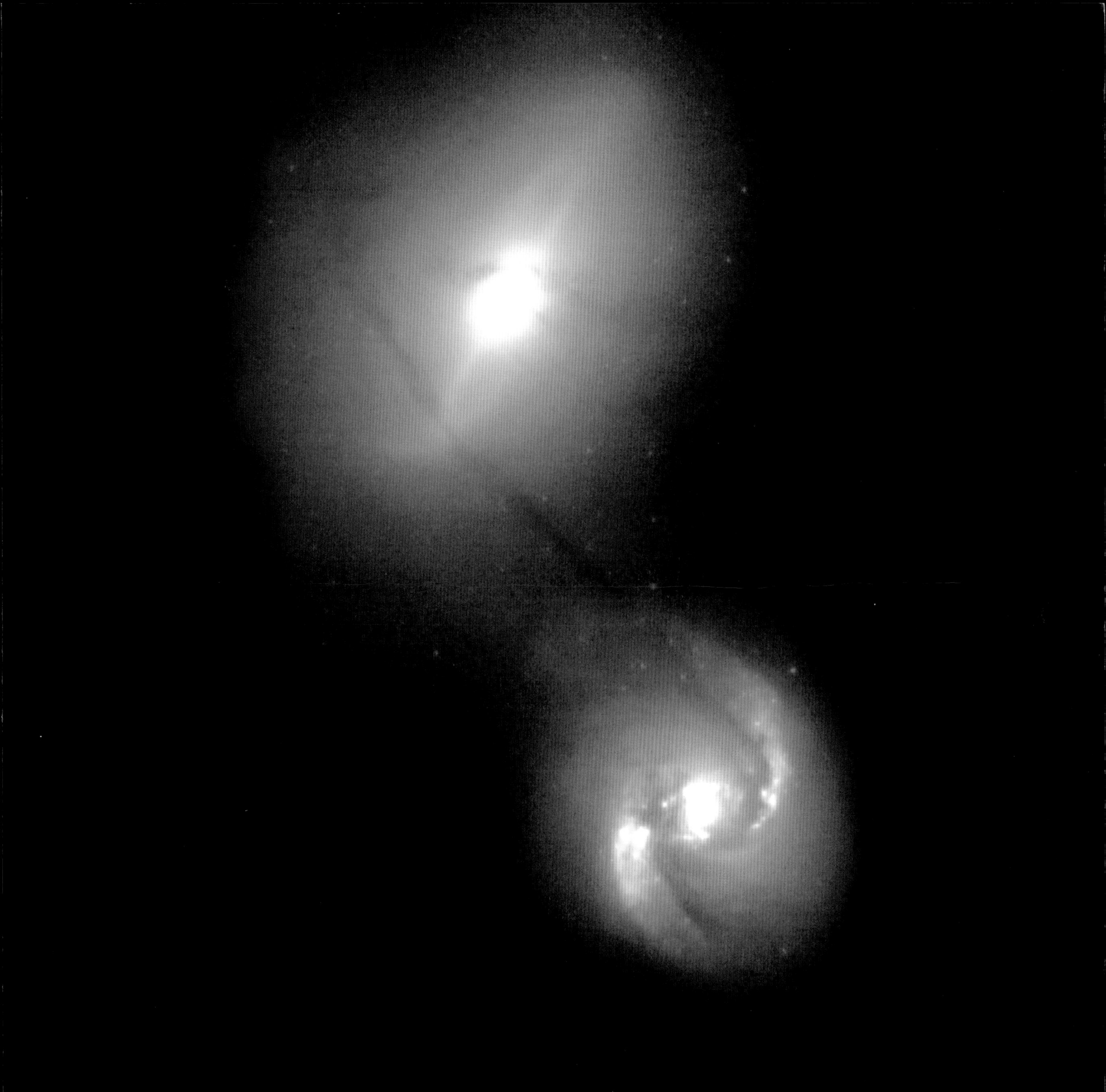

300

million
light years

NGC 1409, NGC 1410

Galactic collision

NGC 1409's gravitational fangs span a gulf of 20,000 light years, creating a dark pipeline that syphons 40 billion trillion tons of matter a year from its smaller companion. Such heavy tonnage actually equates to only two percent of a solar mass, and plainly hasn't been enough to fuel any star formation in NGC 1409 – yet. NGC 1410, on the other hand, simmers with starbursts.

300

million
light years

The Mice

NGC 4676a, 4676b

Galactic collision

The Mice may be named for their trailing tails of stars and dust, but they swoop around each other like two merging flocks of swallows. Two spiral galaxies sculpted by the gravitational shear of their encounter, the Mice are very much a work in progress: it will be billions of years before they finally coalesce into a giant elliptical galaxy – if they can manage to avoid the farmer's wife.

300 **million** light years	*AM 0644-741* Ring galaxy	If the relative velocities of two colliding galaxies are great enough they will tear through each other rather than merge. The wake of such an impact has changed the face of this galaxy, sweeping away its spiral structure as a tsunami of star formation races across its disc. The wave now engulfs 150,000 light years, its crest glowing with the ultraviolet radiation of hypergiant stars.

400 **million** light years	*HCG 87* Galactic cluster	This troupe of four galaxies, known as the Hickson Compact Group 87, performs an intricate dance orchestrated by the ebb and flow of gravity's tides. A graceful minuet compared with some of the feverish galactic tarantellas we have seen, the dance stirs its performers nonetheless: nudging clouds of dust and gas into supermassive black holes and sparking bright rashes of star formation.

420

million
light years

Tadpole Galaxy UGC 10214

Spiral galaxy

Having come off second best in a galactic close encounter, the Tadpole is now blessed with a fine 300,000-light-year tail of stars, gas and dust. The culprit can be glimpsed through its main spiral arms (top left). Like its terrestrial namesake, the Tadpole will loose its tail as it matures – the blue star clusters it contains will become dwarf satellite galaxies, not unlike the Magellanic Clouds.

million
light years

Hoag's Object PGC 54559

Ring galaxy

A nearly perfect smoke ring of hot young blue stars frames a nucleus of older and cooler yellow stars (and, incidentally, another distant ring galaxy). As we have seen, ring galaxies are usually formed in the wake of a high-speed galactic hit-and-run incident. But with no trace of an intruder galaxy, astronomers suspect the outer ring may be the shredded remains of an ex-neighbour.

5 the edge

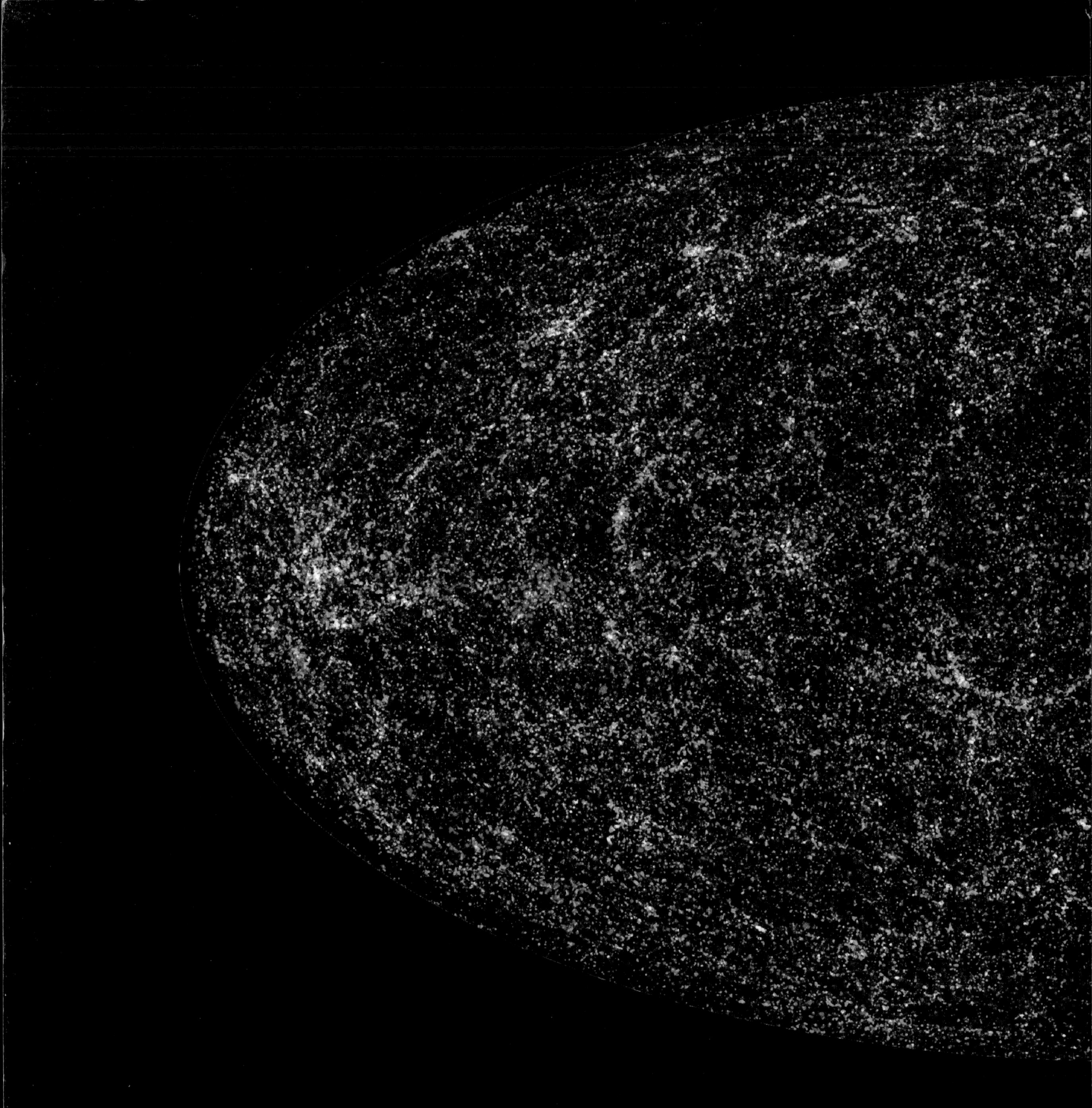

1.3 **billion** light years	*2MASS* {2 Micron All Sky Survey} Local universe map	We are now over a billion light years from home. We have explored planets, stars and entire galaxies but what of the larger view, the universal landscape? Mapping the position of 1.6 million galaxies within 1.3 billion light years of the Milky Way reveals that they cluster like grains of dust on a veil of celestial cobwebs. The nearest galaxies are represented in blue, the most distant in red.

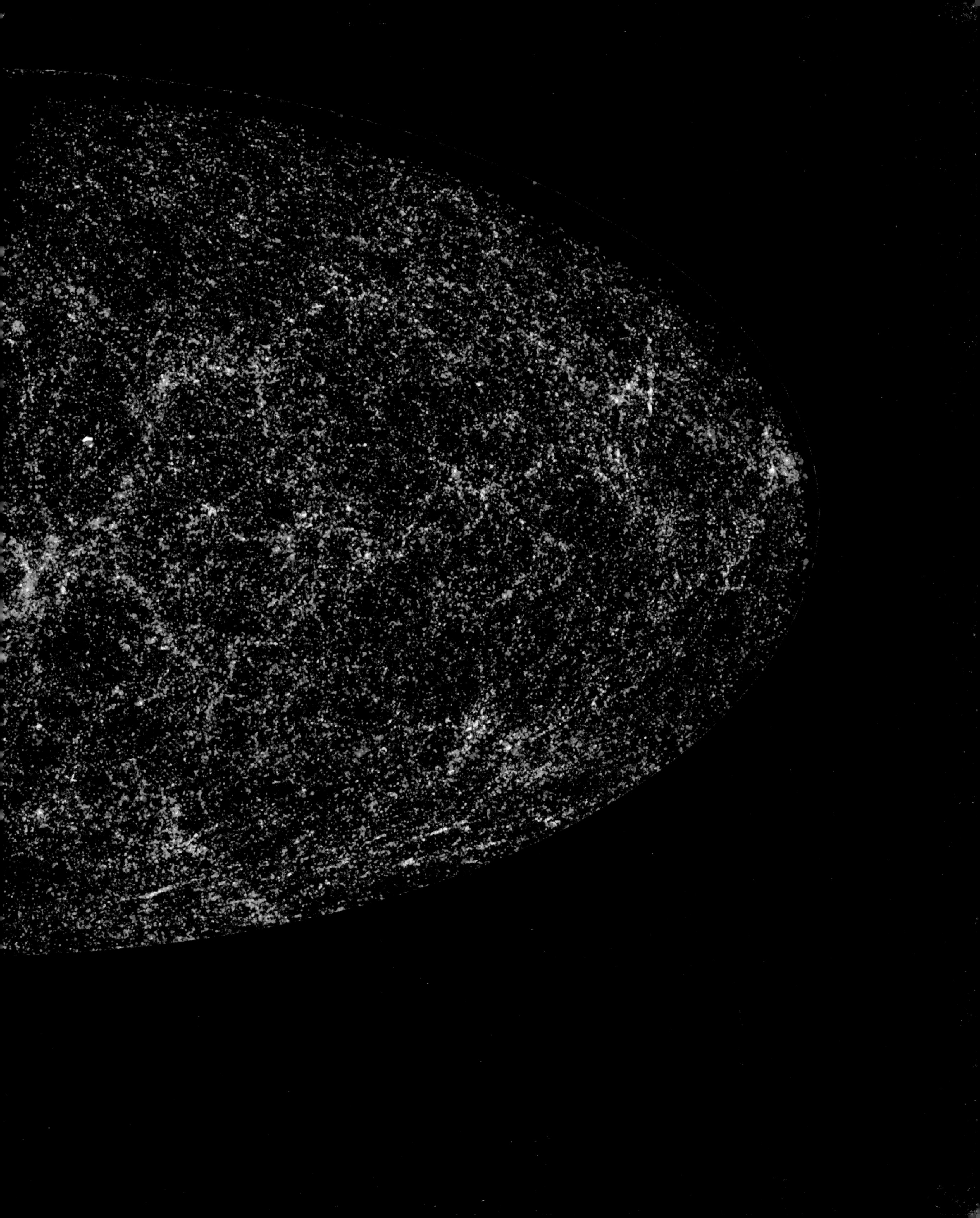

2.2

billion
light years

Abell 1689

Galactic cluster

Hundreds of galaxies crowded by trillions of stars populate a metropolis so massive it visibly warps space, surrounding Abell 1689 with an optical halo of distorted, duplicated and amplified galaxies. Following the strained curve of spacetime, light from more distant objects is 'bent' as it travels through the cluster – a phenomenon known as gravitational lensing.

3 billion light years	*Abell 2218* Galactic cluster	Abell 2218 also magnifies the heavens. Its lens is powerful enough to resolve a galaxy 13 billion light years distant, poised near the edge of space and time. Projected more than once, the galaxy is most easily seen as the smeared dark red dashes on the left side of the cluster. Galaxies this far away were among the first to colonize the universe, bringing an end to the cosmic Dark Ages.

9 **billion** light years	*GRB 990123* Gamma ray burst	For a brief moment on 23 January 1999, a burst of gamma rays equalled the radiance of 100 thousand trillion Suns. Fortunately the blast was two thirds the way to the edge of the universe – if it had been within the Milky Way, it would have sterilized the Earth. Gamma ray bursts are associated with massive supernovae but may also be sparked by the collision of neutron stars or black holes.

11 **billion** light years	*Cloverleaf Quasar* H1413+117 Quasar	A remote species of active galaxy capable of radiating as much energy per second as 1,000 or more galaxies, quasars are the feeding flares of extremely energetic, supermassive black holes. This kaleidoscopic portrait was created in the prism of a gravitational lens, which as well as magnifying our target 50,000 times has duplicated it four-fold.

13

billion
light years

Hubble Ultra Deep Field

Galaxies

The Hubble Ultra Deep Field cuts across 13 billion light years to capture some of the most distant galaxies known, formed when the universe was under 800 million years old. These, the smallest and reddest, are pictured alongside a myriad of 10,000 younger galaxies. Their often unusual and irregular forms chronicle an era when order was just beginning to emerge from chaos.

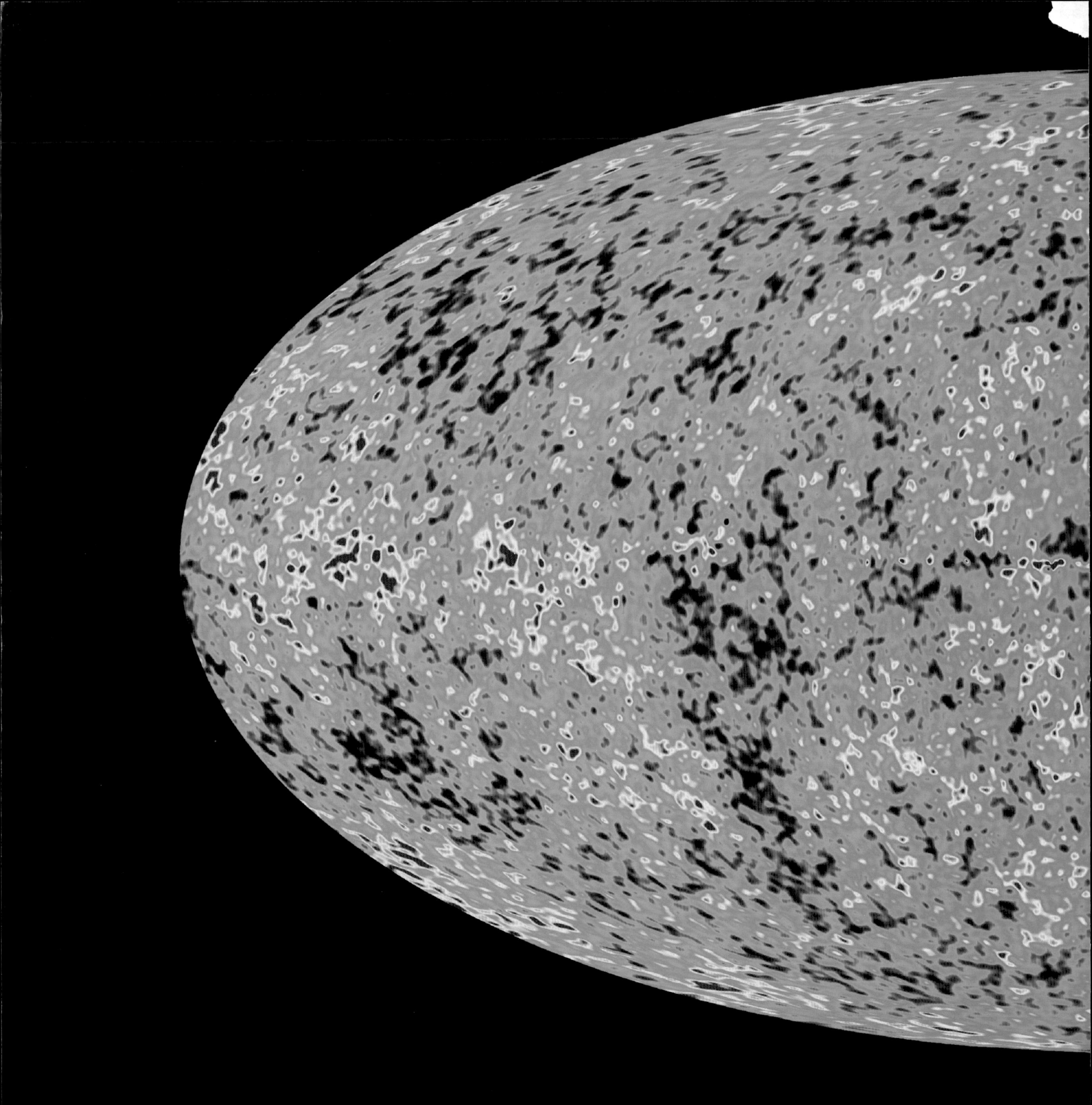

13.4 **billion** light years	*WMAP* Background radiation map	Having closed to within a mere 379,000 years of the Big Bang, a wall of microwave radiation marks the end of our journey, for beyond this horizon an ocean of superheated plasma impedes light's progress. From this primeval vantage point we can discern the overall shape of our cosmic habitat: we may not live on a flat Earth but we do live in a flat universe.

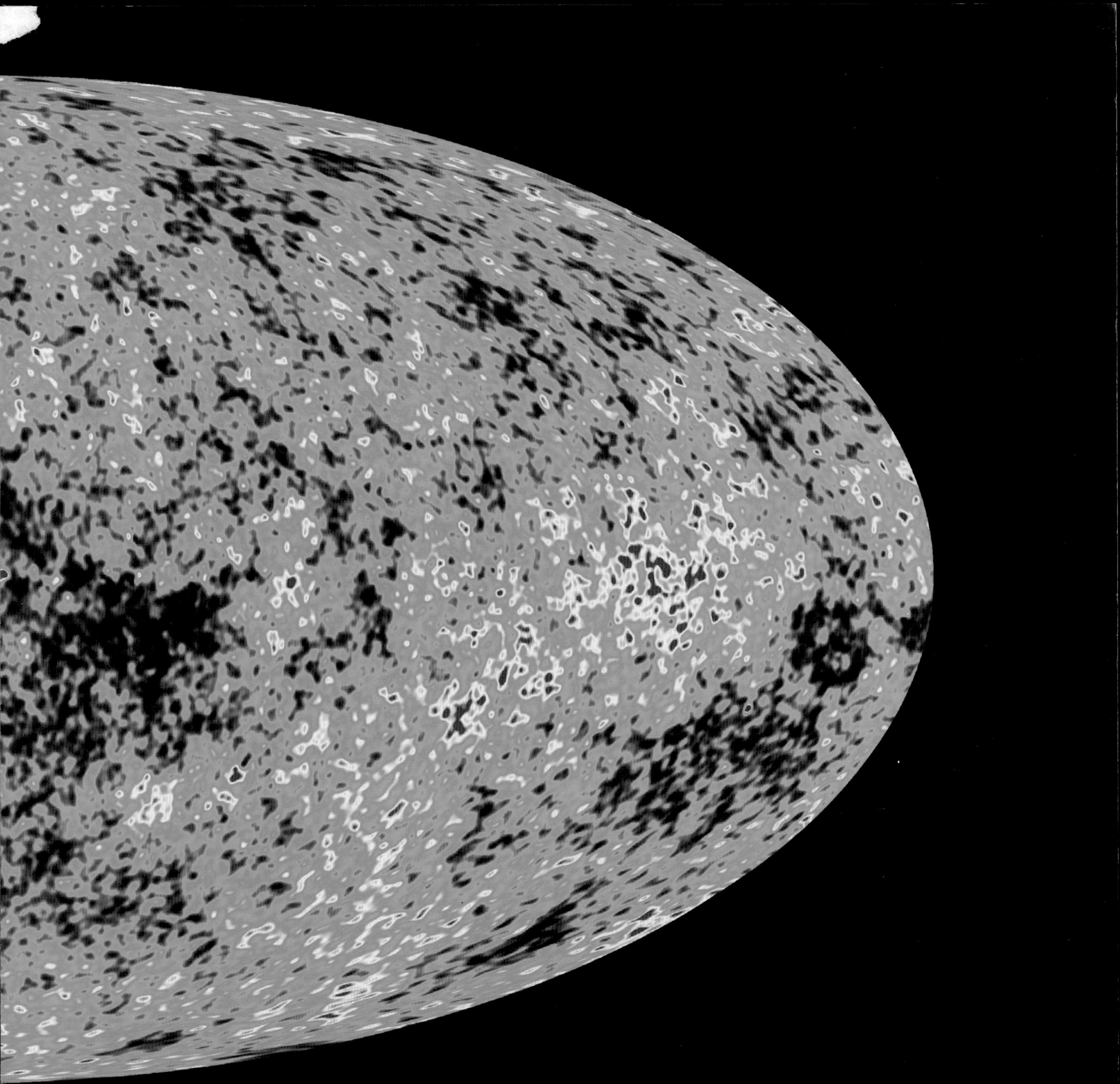

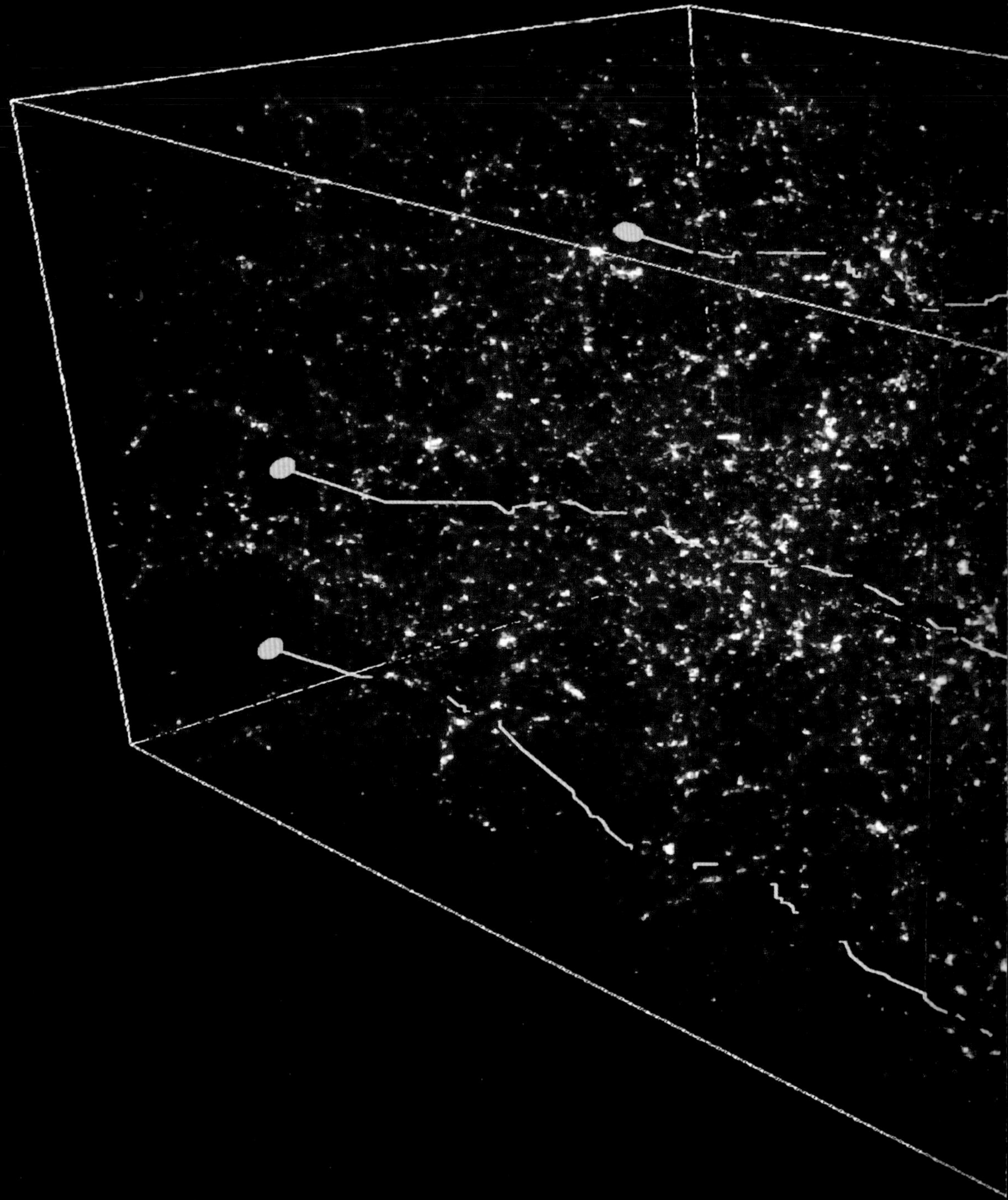

billion
light years

Dark Matter

Dark matter map

There is more to the universe than meets the eye: 87 percent of its bulk is not only invisible to us, but it doesn't even consist of the same stuff as us. Everything – planets, stars and galaxies – we have seen on our 13.4 billion light year journey is but a light froth on a dark ocean. Dark matter is the real stuff of our universe, binding galaxies together and weaving the web to which they cluster.

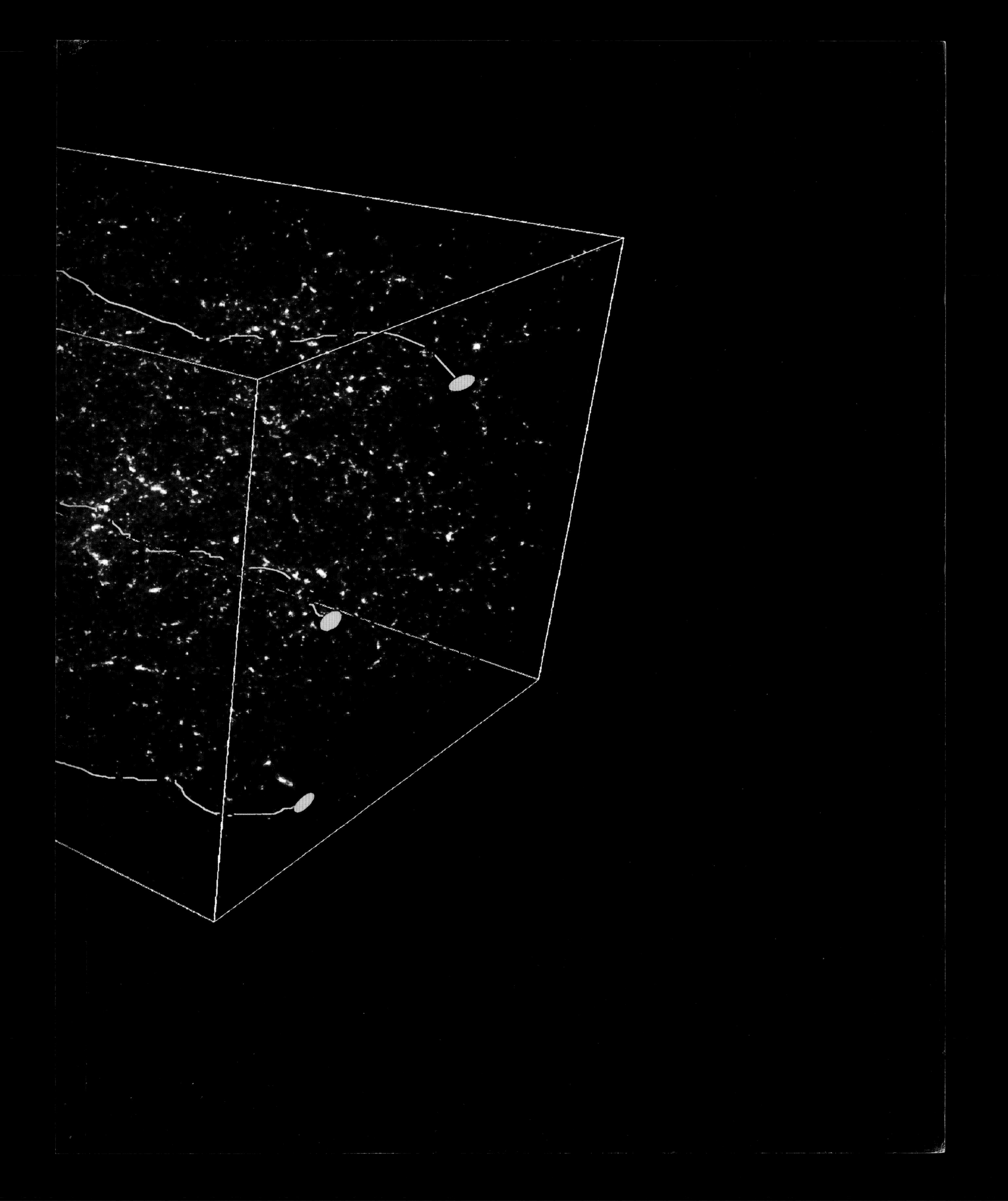

glossary

Absolute zero
The lowest temperature attainable in the universe. Equivalent to –273 °C (-459 °F). There is no upper limit on temperature.

Accretion disc
A ring-like structure formed by material spiralling into a gravitational source, for example a black hole.

Active galaxy
A galaxy that emits more energy than can be accounted for by its normal components: stars, dust and gas.

Active Galactic Nucleus (AGN)
The core of an active galaxy. The energy generated in an AGN may outshine all the other stars in the galaxy. They are believed to be powered by the frictional heating of an accretion disc as it spirals into a supermassive black hole.

Aphelion
Apoapsis in a solar orbit.

Apoapsis
The farthest point in an orbit from the body being orbited.

Asteroid
A small (up to 1,000 kilometres or 600 miles in diameter) rocky body orbiting the Sun. The vast majority are found in the asteroid belt between Mars and Jupiter.

Background radiation see Cosmic background radiation

Big Bang
The moment 13.7 billion years ago when the universe exploded into existence and began expanding.

Binary system
A pair of objects (two stars for example) bound together by mutual gravitation.

Black dwarf
The cold ashes of a Sun-sized star that has evolved into a white dwarf and subsequently cooled to such a degree that it no longer radiates heat. Black dwarfs are theoretical objects as the time taken for a white dwarf to cool is longer than the current age of the universe.

Black hole
A concentration of mass with a gravitational field so strong that – within a certain radius – nothing, not even light, can escape. At the end of their lives, particularly massive stars collapse under their own weight to form black holes.

Bok globule
A dark, dense cloud of dust and gas in which star formation is taking place.

Brown dwarf
Stunted stars with insufficient mass to initiate and sustain hydrogen nuclear fusion.

Collapsar
A generic name for the three types of 'collapsed star' that may form at the end of a star's life: a white dwarf, a neutron star or a black hole.

Comet
A 'dirty snowball' of dust and ice left over from the formation of the solar system. Comets are classified according to their orbital period. 'Short-period' comets complete their orbit in under 200 years, 'long-period' comets can take millions of years. See also Kuiper Belt and Oort Cloud.

Cosmic background radiation
The 'afterglow of creation' or the remnant radiation from the Big Bang. It has now cooled to -270 °C (-454 °F), only 2.7 °C above absolute zero.

Dark energy
A hypothetical form of energy that permeates all of space creating an anti-gravity force that accounts for the observed acceleration of the universe's expansion. Not to be confused with dark matter, whose gravitational effects work in the normal attractive direction.

Dark matter
Matter that cannot be detected by its emitted radiation, although its presence can be inferred by its gravitational interactions with visible matter. Most of the universe's mass exists in this form.

Dark nebula
A cloud of dust and gas dense enough to block the visible light from the objects they obscure. Also known as an absorption nebula.

Dwarf galaxy
A small, faint galaxy, either irregular or elliptical in structure.

Electromagnetic radiation
The most familiar type of electromagnetic radiation is light, but visible light is but a brief archipelago in an ocean of wavelengths. The full electromagnetic spectrum runs from extremely energetic gamma-rays, to low power waves via x-rays, ultraviolet, visible and infrared light and microwaves. Electromagnetic radiation can also be described as a stream of particles known as photons.

Elliptical galaxy
A galaxy that appears spherical or American-football-shaped with no specific internal structure.

Emission nebula
A cloud of gas that shines with its own light. Usually, such nebulae absorb ultraviolet radiation and re-emit it at visible wavelengths. Examples of emission nebulae include planetary nebulae and supernova remnants.

Event horizon
The point of no return surrounding a black hole. Anything crossing this boundary is effectively lost from the universe.

Fluorescence
An optical phenomenon whereby a molecule absorbs an invisible high-energy ultraviolet photon and re-emits it as a visible, lower-energy photon.

Fusion see Nuclear fusion

Galactic cluster
A collection of dozens to thousands of galaxies bound together by gravity.

Galactic halo
A spherical region around a spiral galaxy that contains dim stars and globular clusters. The radius of the halo surrounding the Milky Way extends some 50,000 light years from the galactic centre.

Galaxy
A gravitationally bound system consisting primarily of dust, gas and stars. Galaxies range in scale from hundreds to hundreds of thousands of light years and are classified according to their appearance: Spiral, Elliptical, Lenticular, Ring and Irregular.

Gamma ray
The most energetic form of electromagnetic radiation.

Gamma ray burst
A brief but immensely powerful burst (measured in minutes at most) of gamma rays from space. Believed to be triggered by supernovae or the collision of neutron stars or black holes.

Globular cluster
A distinct, densely packed sphere of stars that can approach a population density one thousand times greater than our stellar neighbourhood. Globular clusters were thought to be relics of the universe's first stellar generations, but recent observations of starburst galaxies have revealed globular clusters in the making.

Gravitational lens
A massive object that magnifies or distorts the light of objects lying behind it.

Gravity
A physical force that appears to exert a mutual attraction between all masses, proportional to the mass and distance of the objects. Einstein's Theory of General Relativity explains gravity as the curvature of spacetime.

Heliopause
The boundary between our solar system and interstellar space – where the pressure of the solar wind equalizes with the interstellar medium. Believed to lie between 14 and 20 light hours from Earth.

Herbig-Haro Object
A distinctive class of emission nebula created by jets of material ejected by newborn stars colliding with the interstellar medium.

Hypergiant
The most massive and luminous stars. The largest examples are over 100 times more massive than our Sun and millions of times more luminous. They burn brightly but briefly – surviving for no more than a few million years. See also O-class star.

Hypernova
An unusually powerful supernova. Associated with the demise of particularly massive stars.

Index Catalogue (IC)
A two-part supplement adding 5,386 astronomical objects to the New General Catalogue (NGC), published in 1895 and 1908.

Interstellar medium
The extremely rarefied 'atmosphere' of space, typically consisting of 90 percent hydrogen, 9 percent helium and 1 percent dust.

Infrared
A section of the electromagnetic spectrum invisible to human eyes, but sensed as heat or thermal radiation.

Ionization
The process that produces ions – atoms that are electrically charged by the capture or loss of electrons. Atoms can be stripped of their electrons by high-energy radiation (from stars, for example). Material that has been completely ionized is known as a plasma.

Irregular galaxy
Any galaxy that lacks the necessary structures to be classified as an elliptical, spiral or lenticular galaxy.

Kuiper Belt
A disc of icy bodies (Kuiper Belt Objects, or KBOs) stretching from Neptune's orbit out to at least one light day from the Sun. Its largest known resident is Pluto.

Lenticular galaxy
A galaxy shaped like a spiral galaxy with a central nucleus and a disc, but lacking spiral arms.

Light
Electromagnetic radiation the human eye can detect. However, the term can also be applied to all electromagnetic radiation.

Light second
The distance covered by light in a vacuum in a second: 299,791 kilometres (186,282 miles)

Light minute
The distance covered by light in a vacuum in a minute: 17.9 million kilometres (11.1 million miles)

Light hour
The distance covered by light in a vacuum in an hour: 1 billion kilometres (620 million miles)

Light year
The distance covered by light in a vacuum in a year: 9.5 trillion kilometres (5.9 trillion miles)

Local Group
A largely gravitationally bound collection of approximately 40 galaxies spread over 10 million light years, dominated by our own Milky Way and the Andromeda Galaxy.

Luminosity
The amount of radiation emitted by a star or celestial object in a given time.

Magnetar
A species of neutron star with an extremely strong magnetic field.

Main sequence
The main sequence is best pictured on a Hertzsprung-Russell diagram, where a star's position corresponds to its luminosity and temperature. When plotted in this fashion, 90 percent of all known stars will fall into a well-defined curving diagonal line known as the main sequence. A star on the main sequence is a healthy star, converting hydrogen into helium by nuclear fusion. Below the main sequence one finds white dwarfs, above it short-lived, unstable supergiants. Stars are not confined to the main sequence throughout their lives. For example, our Sun will slide off the main sequence, above and to the right of its current position, as it exhausts its supplies of hydrogen and swells to become a red giant. Finally as a white dwarf it will sink to the bottom of the graph.

Mass
A measure of the amount of matter in a body. With his famous equation $E=mc^2$, Einstein showed that mass (m) is equivalent to energy (E). In astronomy there is a careful distinction between mass and weight – weight is the force acting on an object with mass in a gravitational field.

Messier Catalogue (M)
A list of about 100 nebulous-looking astronomical objects compiled by Charles Messier between 1758 and 1781.

Milky Way
Our galaxy. The name is derived from our perception of it as a misty band of stars that divides the night sky. The Milky Way is a large spiral galaxy spanning 100,000 light years and containing over 200 billion stars. Our solar system lies about two-thirds of the way towards the edge of its disc.

Molecular cloud
An accumulation of dust and gas markedly more dense than the interstellar medium. Such clouds can span hundreds of light years and are the exclusive sites of star formation.

Moon
The natural satellite of a planet. There are at least 140 moons in the solar system. More specifically, 'Moon' refers to the natural satellite of Earth.

Nebula
A cloud of dust and gas in space. See Molecular cloud, Emission nebula, Reflection nebula and Dark nebula.

Neutron star
An extremely dense stellar remnant produced in a supernova where huge gravitational forces have compressed electrons into protons producing neutrons. See also Pulsar and Magnetar.

New General Catalogue (NGC)
A compilation of 7,800 astronomical objects published in 1888.

Nova
A star that suddenly increases in brightness by a factor of more than a hundred. Novae occur in binary systems where one component is a white dwarf. Material from the companion star is transferred onto the white dwarf triggering explosive nuclear reactions, resulting in the increased brightness.

Nuclear fusion
The process that powers stars. Two atomic nuclei are forced together, forming a single larger nucleus and releasing energy. Most stars convert hydrogen into helium, more massive, hotter stars can fuse heavier elements.

Orbit
The trajectory of one celestial body around another.

Oort Cloud
A swarm of comets thought to surround the Solar System between a light week and a light year from the Sun. Its existence has been deduced from studies of long-period comet orbits, which seem to have their aphelia in this zone.

Open star cluster
A group of young stellar siblings. Born of the same molecular cloud but only loosely bound together by gravity, such clusters are fated to disperse over a period of several hundred million years.

O-class star
A massive star at the hottest extreme of the stellar spectral classification, typically massing between 30 and 100 solar masses. With a surface temperature greater than 30,000 °C (50,000 °F), their light appears bluish-white and they can also be described as blue giants or blue stars.

Periapsis
The point in an orbit nearest to the body being orbited.

Perihelion
Periapsis in a solar orbit.

Photoevaporation
One of the processes that shapes molecular clouds in areas of star formation. Caused by ultraviolet radiation energizing the cloud sufficiently to overcome the gravity that binds it together.

Photon
A particle of light, the quantum unit of the electromagnetic force.

Planet
An object that orbits a star – the name is derived from the Greek *planates*, or 'wanderers'. By tradition, the name is reserved for large bodies and there is some controversy over where the definition of a planet should end. If Pluto was discovered today, it is unlikely it would be classified as a planet. Instead it would be relegated to the league of minor planets, bodies that include asteroids, comets, Kuiper Belt Objects and Oort Cloud Objects.

Planetary nebula
A luminous shell of debris surrounding a dying star. As red giants stutter to the end of their lives, they cast off their outer layers creating an expanding cloud of dust and gas. Eventually, this mass loss exposes the star's core, which, despite no longer supporting nuclear fusion, typically has a temperature of 100,000 °C (180,000 °F). The shell of debris lights up as it fluoresces in the blaze of ultraviolet radiation pouring from the core. Planetary nebulae are transitory phenomena lasting no more than 100 000 years – fading as their central white dwarfs cool. The name 'planetary nebula' is a historical artefact: through the telescopes of early astronomers they looked like planets.

Plasma
The 'fourth state of matter', an electrically conductive mixture of electrons and ions. Many astronomical objects including stars, emission nebulae and accretion discs consist of plasma.

Protoplanetary disc
The disc of dust surrounding a star out of which planets might form.

Protostar
An embryonic star, not yet massive enough to initiate the nuclear fusion of hydrogen.

Pulsar
A rotating neutron star that emits a sweeping beam of high-energy electromagnetic radiation.

Quasar
The very bright, very distant core of an extremely powerful active galaxy. The word 'quasar' is derived from quasi-stellar radio source, so-called because this class of object was first identified through its radio emissions.

Radiation
The transmission of energy or matter. See also Electromagnetic radiation

Radiation pressure

Red dwarf
A small and relatively cool star with a diameter and mass of less than one-third that of the Sun. Red dwarfs comprise the vast majority of stars.

Red giant
An aging star that has exhausted all the hydrogen in its nucleus and burns increasingly heavy elements as fuel to ward off collapse.

Reflection nebula
A cloud of dust that reflects or scatters the light of nearby stars. Reflection nebulae often take on a blue tint as blue light is scattered more efficiently by dust than red light.

Ring galaxy
A galaxy surrounded by a distinct rim of very bright stars. Thought to be formed when spiral galaxies are pierced by a head-on collision with another galaxy.

Satellite
Any body in orbit about another body.

Solar mass
A measure used to express the mass of stars and larger objects. It is equal to the mass of our Sun, or 1.98 trillion trillion tonnes.

Solar wind
A stream of plasma flowing from the Sun at up to 3.2 million kph (2 million mph). It pervades the entire Solar System up to the heliopause.

Spiral galaxy
A typical spiral galaxy has a spherical central bulge of older stars surrounded by a flattened disc containing a spiral pattern of young, hot stars. The Milky Way is a spiral galaxy.

Star
A massive ball of hydrogen and helium bound together by gravity and shining with the light of nuclear fusion. Gravity provides the energy that drives fusion and fusion provides the power that stops the star from further gravitational collapse. This balancing act determines a star's allotted lifespan – the larger a star, the greater its gravity, the greater its gravity the faster it burns and the faster it burns the shorter its life. The most massive hypergiants burn for a brief, if spectacular, few million years while their smaller cousins measure their lives in billions of years. Once a star has exhausted its fuel it is at the mercy of gravity and will collapse (always dramatically, see Planetary nebula and Supernova) into either a white dwarf, a neutron star or a black hole, depending on its mass.

Starburst galaxy
A galaxy experiencing an intense burst of star formation. Most starbursts are precipitated by interaction – or even collision – with other galaxies.

Star formation
The process by which stars gravitationally coalesce from molecular clouds.

Star formation region
A molecular cloud that is in the process of collapse, forming new stars.

Stellar wind
The solar wind emanating from a star other than our Sun.

Sun
The star at the centre of our solar system. Occasionally referred to as Sol.

Supermassive black hole
A black hole with a mass in the range of millions or billions of solar masses. Most, if not all, galaxies are thought to harbour a supermassive black hole at their core.

Supernova
The explosive demise of a star. Supernovae come in two types: Type I and Type II. A Type I supernova accompanies the collapse of an existing white dwarf made unstable by the accretion of material from a nearby stellar companion. A Type II supernova occurs at the end of a massive star's life as it exhausts its nuclear fuel and starts to collapse under its own weight. As the successive layers of the star collapse into each other a shockwave is produced that blasts the outer layers of the star out into space at speeds in excess of 50 million kph (30 million mph). Depending on the size of the star, either a neutron star or a black hole is left at the centre of the conflagration.

Supernova Remnant (SNR)
An expanding shell of dust and gas – the debris of a supernova explosion, mixed together with swept-up interstellar matter.

T-Tauri star
A class of very young, flaring stars on the verge of becoming normal stars fuelled by nuclear fusion.

Ultraviolet radiation
Electromagnetic radiation with a shorter wavelength than violet light.

Universe
Our bubble of space and time and everything it contains.

White dwarf
The dense, cooling ember of a star that has exhausted its nuclear fuel and collapsed under the force of its own gravity. White dwarfdom awaits all but the most massive stars.

Wolf-Rayet star
A massive star (over 25 solar masses) on the verge of supernova. Wolf-Rayets are characterized by a powerful stellar wind that strips off the outer layers of the star, leaving its core exposed.

X-rays
High-energy electromagnetic radiation. Less energetic than gamma rays, but more so than ultraviolet radiation.

Picture credits: p6 Hubble Heritage Team (AURA/ STScI/ NASA); p12 NASA; p13 NASA/JPL-Caltech; p14 NASA/JPL-Caltech; p15 NASA/JPL-Caltech; p16 NASA/JPL-Caltech/MSSS; p17 NASA/JPL-Caltech; p18-19 NASA/JPL-Caltech; p20 SOHO/EIT (ESA & NASA); p21 NASA/TRACE; p22 NASA/JPL-Caltech; p23 NASA/JPL-Caltech; p24 NASA/JPL-Caltech; p25 NASA/JPL-Caltech; p26 NASA/JPL/Space Science Institute; p27 NASA/JPL-Caltech; p28 NASA/JPL-Caltech; p29 NASA/JPL-Caltech; p30 NASA/JPL-Caltech; p31 NASA/JPL-Caltech; p32 NASA/JPL-Caltech/D. Cruikshank (NASA Ames) & J. Stansberry (University of Arizona); p33 NASA/JPL/Space Science Institute; p34-35 NASA/JPL/Space Science Institute; p36 NASA/JPL/Space Science Institute; p37 NASA/JPL-Caltech; p38 NASA/JPL-Caltech; p39 NASA/JPL-Caltech; p40 NASA/JPL/Space Science Institute; p41 NASA/JPL/Space Science Institute; p42 Kenneth Seidelmann, U.S. Naval Observatory, and NASA; p43 NASA/JPL-Caltech; p44 NASA/JPL-Caltech; p45 NASA/JPL-Caltech; p46 NASA/JPL-Caltech; p47 NASA/JPL-Caltech; p48 NASA/JPL-Caltech; p49 NASA/JPL-Caltech; p50 Dr. R. Albrecht, ESA/ESO Space Telescope European Coordinating Facility; NASA; p51 ESA; p52 NASA, ESA and M. Brown (Caltech); p53 NASA, ESA and M. Brown (Caltech); p54 Voyager image: NASA/JPL. Comp: Nico Cheetham; p55 NASA/JPL; p58 NASA/CXC/SAO; p59 NASA and F.M. Walter (State University of New York at Stony Brook); p60 NASA, ESA and AURA/Caltech; p61 NASA/ESA and The Hubble Heritage Team (STScI/AURA), George Herbig and Theodore Simon (University of Hawaii); p62 NASA, NOAO, ESA, the Hubble Helix Nebula Team, M. Meixner (STScI), and T.A. Rector (NRAO); p63 NASA, NOAO, ESA, the Hubble Helix Nebula Team, M. Meixner (STScI), and T.A. Rector (NRAO); p64 European Southern Observatory; p65 European Southern Observatory; p66 NASA and The Hubble Heritage Team (STScI/AURA); p67 NASA and The Hubble Heritage Team (AURA/STScI); p68 European Southern Observatory; p69 Atlas Image courtesy of 2MASS/UMass/IPAC-Caltech/NASA/NSF/G. Kopan, R. Hurt; p70 NASA, C.R. O'Dell and S.K. Wong (Rice University); p71 K.L. Luhman (Harvard-Smithsonian Center for Astrophysics, Cambridge, Mass.); and G. Schneider, E. Young, G. Rieke, A. Cotera, H. Chen, M. Rieke, R. Thompson (Steward Observatory, University of Arizona, Tucson, Ariz.) and NASA/ESA; p72 NASA and The Hubble Heritage Team (STScI/AURA); p73 Atlas Image courtesy of 2MASS/UMass/IPAC-Caltech/NASA/NSF/G. Kopan, R. Hurt; p74 NASA and The Hubble Heritage Team (STScI); p75 T.A.Rector (NOAO/AURA/NSF) and Hubble Heritage Team (STScI/AURA/NASA); p76 European Southern Observatory; p77 J. Morse/STScI, and NASA; p78 NASA and The Hubble Heritage Team (STScI/AURA); p79 Hubble Heritage Team (STScI/AURA/NASA); p80 NASA and The Hubble Heritage Team (STScI/AURA); p81 NASA and The Hubble Heritage Team (STScI/AURA); p82 Romano Corradi, Instituto de Astrofisica de Canarias, Tenerife, Spain; Mario Livio, Space Telescope Science Institute, Baltimore, Md; Ulisse Munari, Osservatorio Astronomico di Padova-Asiago, Italy; Hugo Schwarz, Nordic Optical Telescope, Canarias, Spain; and NASA; p83 NASA and The Hubble Heritage Team (AURA/STScI); p84 Hubble Heritage Team (AURA/STScI/NASA); p85 Bruce Balick (University of Washington), Vincent Icke (Leiden University, The Netherlands), Garrelt Mellema (Stockholm University), and NASA; p86 Bruce Balick (University of Washington), Jason Alexander (University of Washington), Arsen Hajian (U.S. Naval Observatory), Yervant Terzian (Cornell University), Mario Perinotto (University of Florence, Italy), Patrizio Patriarchi (Arcetri Observatory, Italy) and NASA; p87 NASA; ESA; Hans Van Winckel (Catholic University of Leuven, Belgium); and Martin Cohen (University of California, Berkeley); p88 NASA/JPL-Caltech/W. Reach (SSC/Caltech); p89 NASA and The Hubble Heritage Team (AURA/STScI); p90 NASA, H. Ford (JHU), G. Illingworth (UCSC/LO), M.Clampin (STScI), G. Hartig (STScI), the ACS Science Team, and ESA; p91 NASA, J.J. Hester Arizona State University; p92 NASA, ESA, HEIC, and The Hubble Heritage Team (STScI/AURA); p93 NASA, ESA and The Hubble Heritage Team (STScI/AURA); p94 NASA/JPL-Caltech/G. Melnick (Harvard-Smithsonian CfA); p95 NASA/JPL-Caltech/T. Megeath (Harvard-Smithsonian CfA); p96 NASA/JPL-Caltech/S.Carey (Caltech); p97 NASA and The Hubble Heritage Team (AURA/STScI); p98 NASA, ESA and A.Zijlstra (UMIST, Manchester, UK); p99 Garrelt Mellema (Leiden University) et al., HST, ESA, NASA; p100 NASA, Brian D. Moore and J. Hester (Arizona State University); p101 A. Caulet (ST-ECF, ESA) and NASA; p102 ESA / NASA & Valentin Bujarrabal (Observatorio Astronomico Nacional, Spain); p103 R. Sahai and J. Trauger (JPL), NASA/ESA; p104 NASA, Andrew Fruchter and the ERO Team [Sylvia Baggett (STScI), Richard Hook (ST-ECF), Zoltan Levay (STScI)]; p105 European Southern Observatory; p106 NASA, H. Ford (JHU), G. Illingworth (UCSC/LO), M.Clampin (STScI), G. Hartig (STScI), the ACS Science Team, and ESA; p107 NASA, ESA and J. Hester (ASU); p108 NASA and The Hubble Heritage Team (STScI/AURA); p109 NASA/JPL-Caltech/A. Marston (ESTEC/ESA); p110 NASA, The Hubble Heritage Team (STScI/AURA); p111 NASA and The Hubble Heritage Team (STScI/AURA); p112 NASA/CXC/ASU/J. Hester et al. And NASA/HST/ASU/J. Hester et al; p113 NASA, Jeff Hester and Paul Scowen Arizona State University; p114 NASA, Donald Walter (South Carolina State University), Paul Scowen and Brian Moore (Arizona State University); p115 NASA/CXC/SAO; p116 Raghvendra Sahai and John Trauger (JPL), the WFPC2 science team, and NASA; p117 Jon Morse (University of Colorado), and NASA; p118 NASA, The Hubble Heritage Team (AURA/STScI); p119 NASA and the Hubble Heritage Team (AURA/STScI); p120 NASA, ESA, and The Hubble Heritage Team (AURA/STScI); p121 NASA and Jeff Hester (Arizona State University); p122 NASA/CXC/GSFC/U.Hwang et al; p123 NASA/ESA/JHU/R.Sankrit & W.Blair; p124 NASA/JPL-Caltech/University of Wisconsin; p125 Yves Grosdidier (University of Montreal and Observatoire de Strasbourg), Anthony Moffat (Universitie de Montreal), Gilles Joncas (Universite Laval), Agnes Acker (Observatoire de Strasbourg), and NASA; p126 NASA/McGill/V. Kaspi et al; p127 NASA and The Hubble Heritage Team (STScI/AURA); p128 Matt Bobrowsky (Orbital Sciences Corporation) and NASA; p129 Wolfgang Brandner (JPL/IPAC), Eva K. Grebel (Univ. Washington), You-Hua Chu (Univ. Illinois Urbana-Champaign), and NASA; p130-31 NASA and The Hubble Heritage Team (AURA/STScI); p132 NASA, Don Figer, STScI; p133 Don F. Figer (UCLA) and NASA; p134-135 Atlas Image courtesy of 2MASS/UMass/IPAC-Caltech/NASA/NSF/G. Kopan, R. Hurt; p136-137 NASA/UMass/D.Wang et al; p140 NASA and The Hubble Heritage Team (STScI); p141 NASA, ESA, the Hubble Heritage Team (AURA/STScI), and HEIC; p142 NASA and The Hubble Heritage Team (STScI/AURA); p143 European Southern Observatory; p144 NASA, ESA & Mohammad Heydari-Malayeri (Observatoire de Paris, France); p145 NASA/CXC/Rutgers/J.Hughes et al; p146 NASA and The Hubble Heritage Team (STScI/AURA); p147 M. Heydari-Malayeri (Paris Observatory) and NASA/ESA; p148 NASA, ESA, Y. Nazé (University of Liège, Belgium) and Y.-H. Chu (University of Illinois, Urbana); p149 NASA/CXC/SAO; p150 NASA, ESA, and Martino Romaniello (European Southern Observatory, Germany); p151 NASA/JPL-Caltech/V. Gorjian (JPL) & NOAO; p152 NASA, ESA, Mohammad Heydari-Malayeri (Observatoire de Paris, France); p153 NASA/SAO/CXC; p154 Hubble Heritage Team (AURA / STScI / NASA); p155 ESA/NASA, ESO and Danny LaCrue; p156 Hubble Heritage Team (AURA/STScI/NASA); p157 NASA, ESA and A. Nota (STScI/ESA); p158 NASA/CXC/SAO; p159 NASA/CXC/PSU/S.Park et al; p162 NASA, ESA, and The Hubble Heritage Team (STScI/AURA); p163 NASA and The Hubble Heritage Team (STScI/AURA); p164 NASA / Galaxy Evolution Explorer (GALEX); p165 NASA, ESA and T.M. Brown (STScI); p166 NASA and The Hubble Heritage Team (AURA/STScI); p167 ESA, NASA and P. Anders (Göttingen University Galaxy Evolution Group, Germany; p168 Hubble Heritage Team (AURA/STScI/NASA); p169 E.J. Schreier (STScI) and NASA; p170 NASA, ESA, R. de Grijs (Institute of Astronomy, Cambridge, UK); p171 NASA/JPL-Caltech/S. Willner (Harvard-Smithsonian Center for Astrophysics); p172 NASA, Andrew S. Wilson (University of Maryland); Patrick L. Shopbell (Caltech); Chris Simpson (Subaru Telescope); Thaisa Storchi-Bergmann and F. K. B. Barbosa (UFRGS, Brazil); and Martin J. Ward (University of Leicester, U.K.); p173 NASA and The Hubble Heritage Team (STScI); p174 NASA and The Hubble Heritage Team (AURA/STScI); p175 NASA and The Hubble Heritage Team (STScI/AURA); p176-177 NASA and The Hubble Heritage Team (STScI/AURA); p178 NASA and The Hubble Heritage Team (STScI/AURA); p179 NASA, Gerald Cecil (University of North Carolina), Sylvain Veilleux (University of Maryland), Joss Bland-Hawthorn (Anglo- Australian Observatory), and Alex Filippenko (University of California at Berkeley); p180 NASA and The Hubble Heritage Team (STScI/AURA); p181 NASA and Jeffrey Kenney (Yale University); p182 NASA and The Hubble Heritage Team (STScI/AURA); p183 Hubble Heritage Team (AURA/STScI/NASA); p184-185 NASA, ESA, and The Hubble Heritage Team (STScI/AURA) Acknowledgment: P. Knezek (WIYN); p186 Brad Whitmore (STScI) and NASA; p187 NASA, The Hubble Heritage Team and A. Riess (STScI); p188 L. Ferrarese (Johns Hopkins University) and NASA; p189 NASA and The Hubble Heritage Team (STScI/AURA); p190-191 NASA, ESA, and Hubble Heritage Team (STScI); p192 NASA and The Hubble Heritage Team (STScI/AURA); p193 NASA and The Hubble Heritage Team (STScI/AURA); p194 NASA and The Hubble Heritage Team (STScI/AURA); p195 NASA/CXC/SAO; p196 NASA, Jayanne English (University of Manitoba), Sally Hunsberger (Pennsylvania State University), Zolt Levay (Space Telescope Science Institute), Sarah Gallagher (Pennsylvania State University), and Jane Charlton (Pennsylvania State University); p197 NASA, William C. Keel (University of Alabama, Tuscaloosa); p198-199 NASA, H. Ford (JHU), G. Illingworth (UCSC/LO), M.Clampin (STScI), G. Hartig (STScI), the ACS Science Team, and ESA; p200 NASA, ESA, and The Hubble Heritage Team (AURA/STScI); p201 The Hubble Heritage Team (AURA/STScI/NASA); p202 NASA, H. Ford (JHU), G. Illingworth (UCSC/LO), M.Clampin (STScI), G. Hartig (STScI), the ACS Science Team, and ESA; p203 NASA and The Hubble Heritage Team (STScI/AURA); p206-207 2MASS/T. H. Jarrett, J. Carpenter, & R. Hurt; p208 NASA, N. Benitez (JHU), T. Broadhurst (Racah Institute of Physics/The Hebrew University), H. Ford (JHU), M. Clampin (STScI), G. Hartig (STScI), G. Illingworth (UCO/Lick Observatory), the ACS Science Team and ESA; p209 ESA, NASA, J.-P. Kneib (Caltech/Observatoire Midi-Pyrénées) and R. Ellis (Caltech); p210 Andrew Fruchter (STScI) and NASA; p211 NASA/CXC/Penn State/G.Chartas et al; p212-213 NASA, ESA, S. Beckwith (STScI) and the HUDF Team; p214-215 NASA/WMAP Science Team; p216-217 INC Group, S.Colombi (IAP).

Index

For Tabitha

Smith-Davies Publishing
46 Dorset Street
London
W1U 7NB

First published in 2005

Copyright © Smith-Davies Publishing Ltd 2005

All rights reserved. No part of this publication may be reproduced, stored in a retrieval system, or transmitted in any form or by any means, electronic, mechanical, photocopying, recording or otherwise, without the prior permission in writing of the copyright owners.

The picture credits on page 221 constitute an extension to this copyright page.

Smith-Davies Publishing Ltd hereby exclude all liability to the extent permitted by law for any errors or omissions in this book and for any loss, damage or expense (whether direct or indirect) suffered by a third party relying on any information contained in this book.

A catalogue record for this book is available from the British Library.

ISBN 1 905204 00 0

Written and designed by Nicolas Cheetham

Printed in China

DA JUL 3 0 2008